The significant role that crowds and mobs play in modern history has been recognized since the French Revolution, and the efforts to understand their origin and behavior constitute an important, if neglected, part of early psychology. In *Crowds, Psychology, and Politics, 1871–1899,* Jaap van Ginneken explores the fascinating relationship among crowd psychologists and the important events of their day.

Examining the work of five theorists in the late nineteenth century, Jaap van Ginneken traces the history of crowd psychology from its inception to the work of the French physician Le Bon – widely considered to be the founder of the field – just before the turn of the century. Although he was the most popular and influential of the crowd psychologists, Le Bon's work was much influenced by his predecessors and by contemporaries in his field, a debt he never acknowledged. Jaap van Ginneken traces the descendants and heirs of Taine, Sighele, Fournial, Le Bon, and Tarde, using unpublished correspondences to shed new light on their mutual relations.

Crowds, Psychology, and Politics, 1871–1899 also brings together the important events of the nineteenth century and the work being done on crowd psychology, examining the effects that events, such as the Paris Commune revolt and the Dreyfus affair, had on the founders of crowd psychology. The approach of each theorist is placed in the context of the debates of the day, such as the "hypnosis" debate between Charcot and Bernheim in psychiatry and the "imitation" debate between Durkheim and Tarde in sociology. The inability of crowd psychology to establish itself as an academic discipline resulted from its multidisciplinary approach toward popular events, although the work of Le Bon remained influential with twentieth-century politicians ranging from Theodore Roosevelt to Adolf Hitler.

Cambridge Studies in the History of Psychology

GENERAL EDITORS: WILLIAM R. WOODWARD AND MITCHELL G. ASH

Crowds, psychology, and politics 1871–1899

Cambridge Studies in the History of Psychology

This new series provides a publishing forum for outstanding scholarly work in the history of psychology. The creation of the series reflects a growing concentration in this area by historians and philosophers of science, intellectual and cultural historians, and psychologists interested in historical and theoretical issues.

The series is open both to manuscripts dealing with the history of psychological theory and research and to work focusing on the varied social, cultural, and institutional contexts and impacts of psychology. Writing about psychological thinking and research of any period will be considered. In addition to innovative treatments of traditional topics in the field, the editors particularly welcome work that breaks new ground by offering historical considerations of issues such as the linkages of academic and applied psychology with other fields, for example, psychiatry, anthropology, sociology, and psychoanalysis; international, intercultural, or gender-specific differences in psychological theory and research; or the history of psychological research practices. The series will include both single-authored monographs and occasional coherently defined, rigorously edited essay collections.

Also in this series

Constructing the subject: Historical origins of psychological research
KURT DANZIGER
The professionalization of psychology in Nazi Germany
ULFRIED GEUTER (trans. Richard J. Holmes)
Metaphors in the history of psychology
edited by DAVID E. LEARY

Crowds, psychology, and politics 1871–1899

Jaap van Ginneken
University of Amsterdam, The Netherlands

CAMBRIDGE UNIVERSITY PRESS
Cambridge, New York, Melbourne, Madrid, Cape Town, Singapore, São Paulo

Cambridge University Press
The Edinburgh Building, Cambridge CB2 2RU, UK

Published in the United States of America by Cambridge University Press, New York

www.cambridge.org
Information on this title: www.cambridge.org/9780521404181

First published 1992
This digitally printed first paperback version 2006

A catalogue record for this publication is available from the British Library

Library of Congress Cataloguing in Publication data

Ginneken, Jaap van, 1943–
Crowds, psychology, and politics, 1871–1899 / Jaap van Ginneken.
p. cm. – (Cambridge studies in the history of psychology)
Includes bibliographical references and index.
ISBN 0-521-40418-5
1. Crowds – History – 19th century. 2. Mobs – History – 19th century.
3. Collective behavior – Study and teaching – History – 19th century.
4. Social psychology – History – 19th century. I. Title.
II. Series.
HM281.G46 1992
302.3′3 – dc20 91–42797
CIP

ISBN-13 978-0-521-40418-1 hardback
ISBN-10 0-521-40418-5 hardback

ISBN-13 978-0-521-03249-0 paperback
ISBN-10 0-521-03249-0 paperback

Contents

Figures, maps, and tables

Figures

Maps

Tables

Preface

This study evolved over a long period of time – too long, I admit. In a sense, it is even a belated product of the 1960s. As an activist and a journalist, I identified closely with the protest movements of those days. As a student and a junior lecturer in the academic field of "mass psychology," however, I was simultaneously confronted with a long line of largely derogatory theories about such phenomena. Thus arose my interest in taking a closer look at them and in reviewing their historical backgrounds.

My prime source of inspiration for this whole enterprise was my teacher Marten Brouwer. I also received encouragement from Constance van der Maesen and my colleagues at the Baschwitz Institute of the University of Amsterdam, where I lectured intermittently during the seventies and eighties. A "positivist" psychologist by training, however, I was lacking in documentary skills, as the historian Ger Harmsen soon pointed out to me.

If these skills have improved since those early days, this is largely due to my successive colleague historians of psychology: first at the Theoretical Psychology Department of Leyden University (which, though called a department, had to make do without professorial guidance for most of its existence) and then at the Foundations of Psychology Department of the University of Groningen, where Pieter van Strien directs a rapidly growing body of major research projects in this field.

Furthermore, I received useful feedback at annual scientific meetings of relevant international societies. These included the regular history panels at the meetings of the International Society of Political

Psychology, with Bill Stone, Tom Bryder, and others as regular participants; as well as presentations of papers on related subjects at meetings of the American and European Cheiron Societies for the History of the Behavioral and Social Sciences. One group of foreign colleagues with which I had particularly fruitful exchanges, and which has greatly contributed to the historiography of social and crowd psychology itself, consisted of Erika Apfelbaum, Ian Lubek, and Gregory McGuire.

Another encouragement was the help of biographers and heirs of the various founding fathers I studied: Maria Garbari of Trento and Anna Maria Gadda Conti Castellini of Milan for Scipio Sighele; Charles Fournial in Draguignan for Henry Fournial; and in Paris Pierre Duverger for Gustave Le Bon; Jean Milet, Guillaume de Tarde, and Mrs. Paul Bergeret de Tarde for Gabriel Tarde. Furthermore, I received kind assistance at universities in Rome, Turin, Lyons, Paris, and Brussels; at the national libraries in Rome, Florence, and Paris; and at the military hospital of Val de Grâce in Paris (for the Fournial files).

Special thanks go to my colleagues C. de Boer, A. den Boon, and R. Veneboer for technical assistance, to my mother Mrs. B. J. van Ginneken–van de Kasteele and to Ms. A. Sundaram for helping edit the English text, to the historians A. Heering and O. Wedman for reading the manuscript and pointing out some errors. Mitchell Ash and William Woodward carefully reviewed the provisional text for weak points and inconsistencies. Needless to add, however, they cannot be blamed for possible remaining mistakes, which are solely my responsibility.

Last but not least, I have to ask forgiveness to the many friends, family members, and administrative staff at various departments, for having bothered them with an endless succession of drafts. Some told me my continuous postponements were a way of refusing to grow up. They were probably right.

Jaap van Ginneken

Introduction

Saint Raphaël is a minor port on the French Côte d'Azur, halfway between Saint Tropez and Cannes. Off-season its quiet harbor, shady promenade, and tiny beach retain some of their charming character. A century ago Saint Raphaël had just been discovered by writers and artists seeking to flee the bustle of Paris. Guy de Maupassant, who had become increasingly ill and somber as he rose to fortune and fame, anchored his yacht here on a beautiful spring day in the late 1880s, and went ashore. Approaching the church in the center of town, he witnessed a newly wed couple coming out. Onlookers raising themselves on tiptoe to see, he felt forced to follow their example. Instead of feeling charmed by the scene however, he was overwhelmed by disgust: he had become "one of a crowd"!

"I at once experience a curious and unbearable feeling of discomfort, a horrible unnerving sensation, as though I were struggling with all my might against a mysterious and irresistible influence. And in truth, I struggle with the spirit of the mob, which strives to take possession of me," he later wrote in his diary. "The same phenomenon, a surprising one, is produced each time a large number of men are together," he continued. "All these persons, side by side, distinct from each other, of different minds, intelligences, passions, education, beliefs, and prejudices, become suddenly, by the sole fact of their being assembled together, a special being, endowed with a new soul, a new manner of thinking in common, which is the unanalysable resultant of the average of the individual opinions."[1]

1. *Afloat* (1889), entry of April 11, pp. 158 ff.

Guy de Maupassant was an individualist, but as such he was well-attuned to the spirit of his time, of his society, and of his class. His social group felt both fascinated and frightened by crowds. Mob scenes had become a recurrent theme in nineteenth-century European stories, novels, and plays. They figured prominently in *The Heart of Midlothian* (1818) by the British author Walter Scott, *I Promessi Sposi* (1825–27, 1840–42) by the Italian author Alessandro Manzoni, *Nôtre Dame de Paris* (1831) and *Les Misérables* (1862) by the French author Victor Hugo, *Shirley* (1849) by Charlotte Brontë, *Jacob van Artevelde* (1849) by the Belgian author Hendrik Conscience, *Ange Pitou* (1851) by Alexandre Dumas, *Lucien Leuwen* (1855, 1894) by Stendhal, *A Tale of Two Cities* (1859) by Charles Dickens, *Salammbô* (1862) and *L'Éducation Sentimentale* (1869) by Gustave Flaubert, *Vojna i Mir* (War and Peace, 1865/69) by the Russian author Leo Tolstoi, *Felix Holt* (1866) by George Eliot, *Libertà* (1882) and *Mastro Don Gesualdo* (1888) by Giovanni Verga, *L'Horrible* (1884) by Guy de Maupassant, *Germinal* (1885) by Émile Zola, *Die Weber* (1893) by Gerhart Hauptmann, *Majesteit* (Majesty, 1893) and *Wereldvrede* (World Peace, 1895) by the Dutch author Louis Couperus, *Der grüne Kakadu* (1899) by the Austrian writer Arthur Schnitzler, and several others by lesser known national writers. Most of these fictional accounts were dramatizations of historical events. They often included both implicit and explicit observations on crowd behavior, which preceded the emergence of crowd theories as such, and which I have analyzed elsewhere.[2]

The unconventional nature of crowd behavior had of course already been identified much earlier. The recorded history and accumulated literature of all great civilizations since ancient times contain references to mob events, and their potential threat to the established order. Revolutionary movements and postrevolutionary regimes often cultivated a positive image of crowds and masses for some time. But more often, after calm had been restored, a negative image recurred.

Of course the history and literature of past centuries were primarily written by and for the educated elite, whereas crowds and mobs were primarily associated with the popular masses. The former felt threatened by the intermittent eruptions of the latter, and it is not surprising that their judgment was tainted by fear. Furthermore, the members of

2. Rebellious mobs in nineteenth-century European fiction. See van Ginneken 1985c.

the elite tended to perceive themselves as unique individuals, whereas they often perceived their inferiors as an amorphous aggregate. All the same, the quasi-unanimity of their verdict on "the crowd" through the ages is rather remarkable.

If crowd phenomena had played a role in all societies since the dawn of civilization, they became even more prominent in Europe during the hundred years between the revolutionary year 1789 and 1888, the year in which Guy de Maupassant published his diary, *Sur l'Eau* (translated as *Afloat*), quoted above. During that same century European society underwent some of the most radical transformations of its entire history. The population more than doubled, urbanization more than tripled, and the major capitals more than quadrupled in size. Steamships and railway services increased mobility; postal and telegraph services expanded communication. On this basis new information and organization patterns emerged: networks of voluntary associations, corporations, bureaucracies, and nation-states.

Against this background, furthermore, a fourfold revolution had taken place. An economic revolution had drastically reduced the power of feudalism and gradually brought the forces of capitalism into play. A technical revolution had increased agrarian revenue, but most of all provoked an industrial take-off. A social revolution was replacing the prime contradiction between landowners and peasants with one between bourgeois and proletarians. And the political revolutions had ended the sovereignty of absolute monarchs and transferred power to elected assemblies. Within this context both everyday interaction and crowd eruptions profoundly changed in both content and form.

Halfway through the nineteenth century the outline of the modern Western world slowly began to emerge. The United Kingdom and France had been nation-states for some time, but various wars around the 1860s (re)united Germany, Italy, and the United States as their main rivals. Industry, technology, and science continued to grow unabated during the 1870s, but one rather new element was the birth of psychology, sociology, and political science as empirical disciplines.[3] In a sense the 1880s were the celebration of this new world, but also of the need for new methods of management and control. This clearly held for France and Italy, for instance, which shared a common heritage in spite of all their profound differences. They were the first

3. More on this subject in a Dutch book edited by myself and J. Jansz on *Psychologische Praktijken – Een Twintigste Eeuwse Geschiedenis*, ch. 1.

major countries where psychosocial monographs on the crowd were published during the 1890s.

Both had long remained Catholic countries, and the church still played a primordial role in ideological life. The political philosophy of the ruling strata had therefore long continued to invoke religion and the supernatural order of things. The social and national transformations of these decades, however, brought new groups and values to the fore. The political philosophy of the ruling strata increasingly invoked science and the natural order of things. Whereas the progressive liberalism of the first quarters of the nineteenth century had focused on "natural" equality and democratic reform, furthermore, conservative liberals throughout Europe gradually shifted ground during the final quarter of the nineteenth century. The bourgeoisie often sought to repair some of its differences with the aristocracy and to restore a united front against the demands of a growing proletariat. Many contributions to the emerging sciences of man and society therefore tended to stress both the natural inequality of man and meritocratic principles. These ideas came to play a major role in early crowd psychology and elite sociology.[4]

In one major dictionary the first definition of the word crowd is "a large number of persons, especially when collected into a somewhat compact body without order," and a second definition "the great body of the people."[5] Crowd theories have tended to focus on the apparent contrast between everyday behavior and exceptional occurrences such as riots and panics. Even today the scientific reflection on such phenomena is dispersed over various disciplines and approaches.

One psychological approach to crowds in a narrower sense, for instance, centers upon the supposed *deindividuation* taking place in crowds: "a complex hypothesized process in which a series of antecedent social conditions lead to changes in perception of self and others, and thereby to a lowered threshold of normally restrained behavior."[6]

A major sociological approach to crowds in a somewhat wider sense, by contrast, centers on the sudden emergence of new *norms,* that is to say the sudden emergence of a new "common understanding

4. The conservative nature of early European sociology is analyzed by Bramson in *The Political Context of Sociology* and by Nisbet in *The Sociological Tradition.*
5. Webster's *New International Dictionary,* 1971, Vol. 1, p. 544.
6. Zimbardo, The Human Choice, p. 251.

as to what sort of behavior is expected in the situation."[7] Other proponents of the latter approach propose to define the field as the study of *collective dynamics,* dealing with "those patterns of social action that are spontaneous and unstructured inasmuch as they are not organized and not reducible to social structure." That is to say, the patterns are not reducible to "a set of statuses defined by relatively stable relationships that people in various positions have with each other."[8]

In a detailed study of the history of this *collective behavior* "paradigm," however, the Dutch sociologist Boef pointed out a problem. Although both laymen and scientists seem to agree on the special nature of "typical" crowd phenomena, all definitions involve a value judgment as to what is to be considered ab-normal, un-expected, unstructured, or un-stable, Boef says.[9] Thus the field is in a sense the product of an artificial separation of "normal" and "abnormal" collective behavior, in which the deviant aspects are singled out for special attention.

Couch has made an inventory of stereotypes about collective behavior, which often recur in these theories. They involve spontaneity and suggestibility, emotionality and irrationality, mental disturbance and lower class participation, and destruction rather than creation. But he concludes that none of these criteria really proves valid upon closer inspection. "Crowd behavior is distinctive, but to emphasize the 'abnormal' dimensions of crowd behavior appears to be fruitless," he says. "The acting crowd . . . is a social system human beings adopt to take action with reference to other systems. As such it is no more and no less pathological or bizarre than other social systems they have developed."[10]

The field of crowd, mass, or collective psychology, as it was variously called, emerged toward the end of the nineteenth century, primarily in Italy and France. It is true that there had been isolated studies on related subjects before, such as Hecker's German monographs on "dancing manias" (1832), other epidemics, and their mental aspects; and Mackay's British study on "money manias" and other "extraordinary delusions" of popular crowds (1841). Yet their emphasis was

7. Turner and Killian, *Collective Behavior,* p. 22.
8. Lang and Lang, *Collective Dynamics,* pp. 4, 6.
9. Boef, *Van Massapsychologie tot Collectief Gedrag,* p. 19.
10. Couch in Evans, ed., *Readings in Collective Behavior,* ch. 7.

more on description than on analysis, and they did not yet build on the new notions of modern psychology and sociology, which really emerged only around the 1880s.

Although dozens of minor fragments and smaller articles were devoted to crowd theories during subsequent years, only three authors are usually identified as having published the first larger papers or books on the subject. These are the Italian Sighele, whose book *La Folla Delinquente* (the criminal crowd) was published in 1891, the Frenchman Le Bon, whose book *Psychologie des Foules* (*The Crowd*) was published in 1895, and the Frenchman Tarde, who published two major articles in 1892–3 and two more in 1898–9. The first two were included in his book *Essais et Mélanges Sociologiques* (1895) and the latter three in his book *L'Opinion et la Foule* (1901). The successive chapters of this study will all focus on one of these texts, and detail their background and significance. But two more books are included here. I will also examine Taine's multivolume work on *Les Origines de la France Contemporaine,* published between 1875 and 1893, which was a major source of inspiration for all three authors and many of their colleagues. And I will deal with Fournial's long-neglected *Essai sur la Psychologie des Foules* (1892), which proves to be the principal missing link between the three other works.

From the moment I became acquainted with these authors, I felt intrigued by the fact that their books had appeared over such a short span of time, and within such a limited area. Especially intriguing was that they all seemed to have arrived independently at the same conclusion: that crowds are usually highly emotional and irrational, destructive rather than constructive, and in need of strong leadership. Thus I became interested in studying the contexts from which these theories originated, both in an intellectual and a social sense.

I began my research into the origins of the field during the first half of the 1970s, first as a junior lecturer at the Baschwitz Institute (then of mass psychology, public opinion, and propaganda) at the University of Amsterdam, and subsequently with a small research grant from the Dutch Ministry of Education. At the time, most texts on the history of crowd psychology were originally in German. One such text was written by Baschwitz himself, a German economist and journalist who had chosen Dutch exile before the Second World War, and who came to found the mass psychology and mass communications departments at the University of Amsterdam thereafter. His handbook *Du und die Masse* (1938, later translated into Dutch as *Denkend Mens en Menigte,* "rational man and the crowd") included a critical review of early theories, and called for a thorough reappraisal of their political implications. Another relevant text was the Swiss Reiwald's remarkably complete overview of crowd theories in *Vom Geist der Massen*, published immediately after the war. A later, and somewhat more theoretical, appraisal was Fischer's Swiss dissertation *Masse und Vermassung* (1961).

In the midseventies, however, two major English publications came out. One was Giner's elegant historical overview, *Mass Society* (1976). Another was the book edition of Nye's impressive dissertation, *The Origins of Crowd Psychology* (1975). It was a fairly complete analysis of the life and work of Gustave Le Bon, followed in subsequent years by related studies on his relations with Sorel, the elite sociology of Pareto, Mosca, and Michels, and still later by a book on early French criminology. This clearly set an example for a different approach to the history of the field. During the latter half of the seventies, therefore, when living in Paris but occasionally returning to Amsterdam to teach, I traced and visited descendants and archives of some main authors, and wrote a series of papers about them. Successive versions of these were circulated among students and colleagues in mimeographed form.[11]

It was not until the academic year 1980–1, however, that I finally began to write a real manuscript for my dissertation. During the latter year, however, another excellent American study came out: Barrows's *Distorting Mirrors – Visions of the Crowd in Late Nineteenth Century France*. It covered much of the same ground I was working on, and included a number of similar findings. Looking back, this was not entirely surprising: There had been a revival of interest in crowd theories for some time, and if one followed the trail back from the "classical" authors, one was bound to identify some of the same backgrounds. Because Barrows's book was difficult to surpass, I suspended my book project once again and settled for published papers and articles for some time; these covered the background of various early Italian, French, German, Austrian, and British crowd theorists. Meanwhile others, too, published major articles (Geiger 1977, Cochart

11. The 1974–5 mimeographed Dutch version covered some 450 pages and dealt with the emergence of crowd psychology in Italy, France, Austria, Germany, up to and including some present-day American approaches. A later mimeographed English version on Ferri, Sighele, Taine, Zola, Le Bon, Tarde, Nietzsche, and Michels had been reduced to about half that size. The half-completed 1980–1 typewritten Dutch manuscript covered part of the present Chapters 2 through 4.

1982, McGuire 1984 a.f., Métraux 1983, among others) or chapters in books (such as those edited by Graumann and Moscovici in 1986). Furthermore, major studies came out on authors like Sighele (Garbari) and Tarde (Milet, Clark, Lubek).

Yet, some scholars continued to see Le Bon as the sole inventor of crowd psychology. This tendency was particularly marked in three Parisian publications of the 1980s. It began with a book by one of the most well-known French social psychologists of the day, Moscovici. His *L'Âge des Foules* (1981) was a passionate plea for a reconsideration of Le Bon's ideas on the crowd, along with some of those of Tarde and those of Freud. He definitely had a point, but tended to overstate it. In doing so he made a considerable number of mistakes, and quite a few of these remained uncorrected in the English translation, *The Age of the Crowd – A Historical Treatise on Mass Psychology,* which came out four years later (see the reviews by Thiec and Tréanton 1983, McGuire 1986b, among others).

A second Parisian study was a *thèse de troisième cycle* titled "Sociologie et Lecture de l'Histoire chez Gustave Le Bon" (1982) by Vlach, supervised by one the most well-known French sociologists of that day, Aron, but reproduced only in mimeographed form. And the third study completing this trend was Rouvier's *Les Idées Politiques de Gustave Le Bon* (1986), for which she received a prize from her Parisian university. It was published by the prestigious Presses Universitaires de France. These two studies both claimed to give a reliable summary of all the ideas Le Bon had put forward in his thirty to forty books, suggested that he had invented crowd psychology and a host of other disciplines almost singlehandedly, but had been unjustly ostracized by subsequent generations of social scientists. They ignored or misrepresented many aspects of Le Bon's works that did not fit into their presentation of him as a lonely hero, and also ignored or misrepresented the work of many relevant predecessors, contemporaries, and even followers.[12] A few examples will illustrate this point.

All three books misrepresent the works of Le Bon's Italian predecessors in various ways.[13] They completely ignore the major "missing

12. Van Ginneken, "De constructie van de mythe van de eenzame held – Het geval Le Bon," *Psychologie & Maatschappij* No. 48 (1989): 253–65.

13. A considerable number of articles on the crowd had already been published in Italy (see Chapter 2). The authors unquestioningly adopt Le Bon's own claim that they considered the crowd in purely negative terms (Moscovici 1981, p. 109; Vlach, p. 252; Rouvier, p. 92). This

link" between the Italian and French authors, that is to say Fournial's book on crowd psychology, *Psychologie des Foules,* which preceded Le Bon's by more than three full years.[14] They consistently garble the names of authors, the titles of books, and the years of publication of related texts in English[15] and German.[16] The later authors even copy such mistakes from earlier ones.[17] Furthermore they exaggerate both the disregard of Le Bon at home and his recognition abroad to make their point.[18] They tone down his extremist views to build him into a

is too simple: In fact they did hardly more so than Le Bon himself (in spite of his claims to the contrary). Moscovici also copies Le Bon's claim that the Italian authors wrote about the crowd only in juridical terms (p. 77). This is again too simple: Almost every single psychological idea in Le Bon's book can be traced to the works of previous authors, including those of the Italians. Rouvier predates Sighele's *La Delinquenza Settaria* by twenty years (p. 85), consistently refers to Sieghele instead of Sighele, wrongly identifies him as a representative of the Right, and a future inspirator of fascism. The latter (rather common) mistake had previously been made by Vlach (p. 116). She also claimed that Le Bon inspired Mosca (p. 257). But their major works appeared in the same year, and she gives no evidence to back up that claim.

14. This is all the more curious because he was mentioned in Tarde's first essay on the crowd, and in the monographs by Nye and Barrows.
15. A sampler limited to Anglo-American psychosociology and social psychology. Rouvier gives Park's German dissertation a French title in an American edition, and has it published seventeen years too early (p. 272). She makes Barnes's famous book *From Lore to Science* into one titled *From Lobe to Science* (sic). Moscovici rebaptizes the famous social psychologist Ross Ron (p. 82). Vlach calls the famous social psychologist McDougall "a convinced democrat" (p. 248), which is ignoring his later American works. They garble the name of Burgess into Brugess, Cantril into Cantrill, Hobsbawm into Hobbawm, Katz into Kats, Mosse into Mossé, Smelser into Smelner, etc.
16. A sampler limited to Austro-German depth psychology. Rouvier claims that Le Bon acknowledged the work of Freud in 1895 and 1912. This is unlikely, and she gives no evidence to back this claim (pp. 64, 99). Moscovici garbles the name of the Jungian Reiwald into Rewald and Reinwald, no less than five lines apart (p. 96). Rouvier has Reich's *Massenpsychologie* published in Munich in 1933, which is absurd for a number of reasons. Moscovici makes the Austrian Broch a German. Rouvier makes his book on *Der Massenwahn Theorie* into one on *Der Massenmann Theorie* (sic), and so on.
17. Rouvier, for one, copies the misspelling of Grunenberg, Reiwald, and others from Moscovici, and also some of the false claims referred to before (e.g., those on the Italians).
18. At one point Rouvier acknowledges that Le Bon's book was reprinted eighteen times before 1910, but at another she maintains that he became only "truly successful" between that year and 1930 (pp. 17, 86). Vlach even says that Le Bon was virtually ignored until after his death (p. 90). Moscovici claims that French publications during the fifty years after his death "never" mentioned his extraordinary influence on the social sciences (p. 73), which is a gross exaggeration. He goes on to claim that apart from Tocqueville and Sorel no single Frenchman exerted such an influence abroad. He refers back to Nye's authoritative monograph on *Le Bon and The Crisis of Mass Democracy* to back up this claim. It supposedly said that his success was so great, that "no single other social thinker could rival with him" (p. 79). But he left out two important qualifications: "perhaps" and "in this period" (compare Nye 1975, p. 3). Rouvier had this same study deal with "the crisis of man democracy"

humanist and liberal visionary.[19] In his laudatory preface to Rouvier's book, former prime minister Faure even acclaims his hero as "relativist, inclined to modesty, open to tolerance," which is nearly the exact opposite of what we know about his personality from contemporaries and documents.[20]

To my relief, then, there was apparently still room for yet another monograph on the true origins of crowd psychology.[21] What this study sets out to do, then, is to correct some of the mistakes that have gradually crept into some standard accounts, and have since been repeated again and again. Three sources of bias stand out in particular: an undue emphasis on the contributions from one major language area, on those of a few well-known figures, and on a monodisciplinary perspective.

The first source of bias is an undue emphasis on the French role – though I have not entirely escaped this tendency myself. It is of course true that French was (and remains today) much more of an international language than Italian. This implies that French authors tended to exert greater influence, and also that foreign specialists today find it easier to understand them in their original context. Furthermore, most of the early Italian authors belonged to a school that became discredited for both scientific and ideological reasons. Yet this should not lead us to a consistent minimization or misrepresentation of their contributions. Thus many of the articles or books on the origins of crowd psychology get wrong at least one of the central facts concerning

(sic, p. 270). She invokes Schumpeter's famous book *Capitalism, Socialism and Democracy*, which supposedly said that Le Bon had been if not the "founder," then at least the "first theoretician" of crowd psychology (p. 26). But he wrote in fact that Le Bon was "the first effective exponent" – that is to say, the first who had effect, who made an impact. Schumpeter also adds qualifications on the limitations of the factual basis for Le Bon's conclusions, and on his one-sided emphasis on certain negative aspects of aggregation, which Rouvier is careful to omit (compare Schumpeter 1973–75, p. 257).

19. Chapters 4 and 5 will show that, on the contrary, he was a self-proclaimed antihumanist and antiliberal, but most of the relevant fragments (illustrating his racism and antisemitism, for instance) have been conveniently overlooked by the authors.

20. This does not mean, of course, that he did not have a circle of admirers willing to pay him such compliments.

21. After this manuscript was completed and first presented, J. S. McClelland published his study *The Crowd and the Mob – From Plato to Canetti* (London: Unwin Hyman, 1989). As the subtitle indicates, the author covers a much larger period. The three chapters on Taine, Sighele, Tarde, and Le Bon (Fournial is mentioned only in passing) focus on their ideas, deal with some of the relevant background, but cannot delve into them very extensively. After some hesitation, I have decided to refrain from adding a discussion of our points of agreement and disagreement here.

Sighele's monograph: the publication year, the title, the author, or his political affiliations.[22] Even a major recent Italian study on "a hundred years of collective psychology" fails to recognize Sighele's primacy, by predating Tarde's first paper on the subject by no less than four years.[23]

The second source of bias is an undue emphasis on big names, which once again I have not entirely been able to escape. Thus, there is usually an excessive emphasis on Le Bon, somewhat less on Tarde, still less on Sighele. I have focused on their main works, too, but only in order to demonstrate that their books were embedded in a steady stream of articles and papers by contemporaries, and cannot correctly be understood in isolation. Furthermore I hope to show that apart from the three well-known books on crowd psychology there was a fourth that is ignored by most present-day authors; though the book is mentioned by Nye and described by Barrows, its context is only revealed here. Although Fournial's contribution was not particularly influential, it did play a major role in the relations among the other three authors, as we will see.

The third source of bias that I perceive in some of the historiography on the origins of crowd psychology is a certain discipline-centrism.[24] This does not hold for Nye and Barrows, who are both professional historians of the late nineteenth-century intellectual climate in France. But it does hold for some of the psychosocial scientists who went in search of the roots of this field. Thus the psychologist Moscovici (1981) largely views Le Bon and Tarde as forerunners of Freud's psychoanalytic approach. He mentions the sociologist Park but gives no sign of being aware of the alternative approach to "collective behavior" formulated by Park and his pupils. Neither Blumer, nor Turner

22. The 1899 English translation of Le Bon's *Psychology of Socialism* has Sighele's book mistitled *The Criminal Masses* instead of *The Criminal Crowd* (p. 100, n. 1), and twisted its meaning. Park (1904, 1972, pp. 6–7) confuses *La Folla Delinquente* (1891) and the subsequent *La Coppia Criminale*. Reiwald (1946) does not mention the publication year of the first edition of Sighele's book, but only that of the second French edition (1898). Fischer (1961) says Sighele's book was published in 1893 and called *L'Anima* [Animo] *della Folla*, which was in fact Rossi's title of 1898. Giner (1976, pp. 63, 70, 278) mentions *La Folla Delinquente* of 1891, but only refers to *L'Intelligenza della Folla* of 1931. Boef (1984, pp. 73–4) refers to an 1891 book by Sighele but has *La Folla Delinquente* published in 1894–95. Etc.

23. Mucchi-Faina, *L'Abbraccio della Folla*, pp. 24, 28, 280.

24. It should be added that more often than not, major scientific developments result from the migration of ideas from one (sub)discipline or one (sub)culture to another. Monodisciplinary histories tend to misperceive the nature and significance of such movements.

and Killian, nor Lang and Lang, nor any other related author appears in the references. By contrast, the sociologist Boef (1984) largely approaches Le Bon (instead of Tarde) as a forerunner of Park and his pupils. He mentions the psychologist Festinger, but in turn gives no sign of being aware of the alternative approach to "deindividuation" formulated by Festinger and his pupils. Neither Festinger, nor Zimbardo, Diener, nor any other related author appear in the references.

Though my own text occasionally duplicates some of the aforementioned works, I believe it adds more of a multicultural, multiauthor, and multidisciplinary perspective. This holds least for the preliminary chapter on Taine, which is not about crowd psychology proper, but about its major source of inspiration. Compared to the excellent studies by Nye and Barrows, my chapters on Sighele and the Italian authors, and on Fournial and the priority debate, contain a number of fairly new elements, and thereby help to place subsequent authors in an entirely different light. My chapters on Le Bon and Tarde, by contrast, contain few novel discoveries per se and concentrate instead on a further elaboration of both their scientific and ideological background and significance. This brings to light, however, some major lines of influence that have been largely ignored so far.

If the Italians and the French were the first to write psychosocial monographs on the crowd, furthermore, they were not the only ones. I have already published a number of articles on subsequent monographs in English and German by authors such as Trotter, Freud, Reich, and the Frankfurt School, on their social and intellectual context.[25] Building on this material, I hope to find the opportunity to write a sequel, *Crowds, Psychology, and Politics, 1900–1933,* one day.

Concerning these various relevant contexts, though, a word of caution may be in order. The chapters deal with the crowd theories of successive authors, their respective background and impact. They do not deal exhaustively with all the other aspects of this background. Thus, the sections in this book on major historical events such as the Commune Revolt, the rise and fall of General Boulanger or the Dreyfus affair, for example, are necessarily somewhat sketchy. This is a study of intellectual, not social history.

This book is, first of all, a study of the history of psychosocial science.[26] Studies of this kind used to focus on Great Men and Great

25. See the bibliography.

26. I prefer the term *psychosocial science* because the overly casual distinction between mental and social processes in everyday scientific language is dubious and places a heavy burden on both social psychology and collective behavior studies.

Schools: on their gradual discovery of the Universal Laws underlying psychosocial phenomena, on the steady progression of our scientific knowledge about True Reality. Through the works of Weber and Mannheim, Merton and Kuhn, Bachelard and Foucault, however, a different approach has gradually evolved. We now see that science is to a large extent culturally relative. It is not only linked to a personal history but also to a social group, to an *Ort- und Zeitgeist.*

Basically, scientific research is just another way of "making sense": comparable to other ways (religion, education, journalism), but with its own specific rules. Making sense is by definition not only a cognitive but also a social process: Every author necessarily shares certain preconceptions about the world with his social group and contributes to recreating them. Social change and current events place certain problems on the intellectual agenda as insufficiently understood, in need of better explanation, and of eventual control. As a result, current concepts and methodology are mobilized for this endeavor, and a view considered appropriate is constructed through a long and laborious process. After the industrial revolution, for instance, the contrast between normality and abnormality was exacerbated through a large variety of mechanisms.[27] Various forms of deviance were singled out for special attention. This held not only for collective behavior, but for individual behavior, too. Thus, medical doctors generated *The Anatomy of Madness,* for instance – as W. F. Bynum, R. Porter, and M. Sheperd labeled a collection of *Essays in the History of Psychiatry* that they edited.[28]

With the advance of the division of labor, furthermore, increasing attention was given to individual mental capacity. In his book subtitled *Phrenology and the Organization of Consent in Nineteenth-Century Britain,* Roger Cooter convincingly demonstrated *The Cultural Meaning of Popular Science.* He concluded that phrenology "more than any other form of psychology, established in the public mind the notion that human behavior was capable of classification and measure."[29] Gradually, the idea arose that individual mental capacity could and should be measured, and that replacing verbal interpretation with statistical calculation would make it value-free, too. Stephen Jay Gould has shown, however, that from craniometry to IQ-testing this per-

27. See J. van Ginneken, "De psychologische mens – Menswetenschappen en maatschappij rond de eeuwwisseling," ch. 1 in J. van Ginneken and J. Jansz, eds. (1986).
28. 2 vols. (London: Tavistock, 1985).
29. *The Cultural Meaning of Popular Science* (Cambridge: Cambridge Univ. Press, 1984), p. 271.

sistent urge at more elaborate quantification has repeatedly led to the Mismeasure of Man (as his 1981 book title terms it). Because internal processes have all too easily been equated with certain external signs, selected arbitrarily.

Thus, most scientific facts are not as unproblematical as they may seem at first sight. Etymologically, the word *fact* stems from the Latin *facere*, "making." Facts are always artifacts. They are man-made, and the result of the conceptualization, organization, and instrumentation of observation. Attempts at making psychology less qualitative-interpretative and more quantitative-empirical have consistently been haunted by this fate. A series of studies titled *The Rise of Experimentation in American Psychology*, edited by Jill Morawski, has for instance dissected some of the best known "classical" research projects in this field, and found that underneath all their seeming exactitude a large number of arbitrary decisions lay hidden from view.[30]

It was long thought that if only the psychosocial sciences could copy the methods of the natural sciences, this would solve these problems, whereby the researcher and the object of his or her study are interrelated in numerous ways. Today, however, a reverse tendency can be perceived. Scholars of natural science become increasingly aware that even the most "objective" approach always contains subjective elements (and have gradually even called into question this whole contradistinction).

Thomas Kuhn himself has devoted various case studies to the context of the Copernican revolution, to the origins of modern thermodynamics and quantum mechanics. Barry Barnes has developed the social scientific implications of these works.[31] In *Leviathan and the Air Pump*, Steven Shapin and Simon Schaffer have linked the origins of the experimental laboratory itself to ideological considerations in the Hobbes–Boyle controversy. This further served to illustrate, that "solutions to the problem of knowledge are solutions to the problem of social order."[32] Today there hardly remains a major scientific discovery that has not been reviewed in the light of the specific context in which it occurred.

30. New Haven, Conn.: Yale Univ. Press 1988. Reviewed in J. van Ginneken, "Het experiment als bron van artefakten," *Kennis & Methode*, 13, No. 4: 404–13.
31. B. Barnes, *T. S. Kuhn and Social Science* (New York: Columbia Univ. Press, 1982). Also see B. Barnes, *Scientific Knowledge and Sociological Theory* (London: Routledge & Kegan Paul, 1974); and B. Barnes, *Interests and the Growth of Knowledge* (London: Routledge & Kegan Paul, 1977).
32. Princeton, N.J.: Princeton Univ. Press, 1985.

The experimental laboratory, the research institute, the university structure, the academic world, all are transected by extrascientific forces. Furthermore, there usually is a vivid contrast between the way in which scientists claim they reached their conclusions and the way in which they actually did. This holds even for present-day "advanced science." Thus around 1980, two European studies of the daily life and work in advanced biological and chemical research laboratories on the American west coast caused a "revolution" in science studies. The first study, *Laboratory Life – The Construction of Scientific Facts* (1979) was done by Bruno Latour, and written with the help of Steve Woolgar. The second study, *The Manufacture of Knowledge – An Essay on the Constructivist and Contextual Nature of Knowledge* (1981), was written by Karen Knorr-Cetina.[33] Although they differed in intellectual style, both studies used an ethnomethodological approach. That is to say, they looked at the way in which advanced scientists generated "shared meanings" through everyday routines – just as an anthropologist would investigate an unknown tribe and its worldview.

In his more recent work *Science in Action,* the same Bruno Latour has proposed to look at "facts in the making" rather than at the final results, and "without preconceptions as to what constitutes knowledge."[34] Even if crowd psychology dates from the very beginning of psychosocial science and before the emergence of its present-day methodology, this historical study has similar concerns. It is not meant to evaluate the scientificity of the theories in question, or to decide which one provides a better explanation, although I occasionally express a personal preference for certain conceptualizations over others. The book is an exploration of the links between these theories and their various contexts. It turns out that their explanations are very much rooted in the ideas and events of their day.

Where today's psychologists speak of deindividuation, for instance, Le Bon and his contemporaries spoke of hypnotic suggestion. Where today's sociologists speak of interaction, Tarde spoke of imitation. They dealt with similar phenomena, but their conceptualization was attuned to the spirit of their times. So-called Whig histories of science tend to identify these changes automatically with progress, and tend to treat the past as simply as a long and laborious preparation of the

33. The former published by Sage and republished by Princeton Univ. Press in 1986; the latter published by Pergamon, Oxford.
34. London: Open Univ. Press, 1987, pp. 13, 15.

present. Constructivist approaches, by contrast, attempt to replace past theories in their original context, and to understand what they really meant at the time.

Of course there are many ways in which the facts considered relevant to the history of a field might be ordered and presented. I have chosen to focus on the first major books containing modern crowd theories, and to group each chapter around one such book and its author. The book is the prime element to be understood: It was written and published at a specific moment in the life and work of the author, and played a role within the various groups to which he belonged – scientific, ideological, etcetera. Thus, I will begin each chapter with one or two pages introducing the setting, the author, the book, and its central themes. After that, I will go back in time but will have to make distinctions among various authors.

Some of these authors (namely Taine, Le Bon, and Tarde), were already well established scientists when they wrote these works. In those cases it is necessary to begin with a section on their early life and work, covering the years preceding their respective monographs on crowd behavior, that is. This gives a general impression of the relevant aspects of their personality and thought. Other authors, however (namely Sighele and Fournial), had only just begun their professional careers when they wrote their monographs on crowd behavior. In those cases, I will skip the separate section on their previous life and work, and turn first to the broader background.

This relevant background is of a twofold nature. Every science has a material object: It concerns itself with something – in this case crowd behavior. But every science also has a formal object: It has a way of looking at things – for example a psychological approach. If we want to know what inspired the choice of subject matter and the specific approach of such a book, therefore, there are at least two aspects of its background to be dealt with. On the one hand are the vicissitudes of the material object during the years preceding the monograph. That is to say: What was the social situation? Did crowd events play a marked role? What aspects were relatively new and conspicuous? And how did the author and his social group feel about them? In this regard, it is necessary to take a closer look at changing patterns of collective violence, for instance, as studied by Hobsbawn, Rudé, the Tillys, and others. Thus the significance of crowd events changed notably when new forms of political organization arose. Therefore each chapter will contain a section on this social context.

On the other hand are the vicissitudes of the formal object: the constantly evolving scientific approaches. Although a division of labor within the psychosocial sciences was just beginning in the late nineteenth century, there was always a specific cluster of related disciplines in which the author in question was most active, and to which he contributed the most. These disciplines were characterized by current debates on methods, theories, and concepts. In this regard, it is necessary to take a closer look at the history of psychology, for instance, or at the history of sociology.[35]

In this regard, Thomas Kuhn's idea of paradigm shifts springs to mind, of course. It has justly been remarked, however, that this idea cannot easily be transplanted from the natural sciences to the humanities. Within each psychosocial discipline contrasting paradigms coexist even today. The adherence of individual scientists is often partial and multiple, and a shift is never abrupt or total, in the sense of a scientific revolution taking place. Yet, one may well try to identify more or less dominant paradigms within more or less specific groups of scientists, and the gradual emergence of a forceful challenge to them at a particular time and place. Thus, every chapter of this study will also contain a section on this intellectual context.

This is not intended to diminish in any way the originality of these contributions to crowd psychology, or to pretend they are simply "products of their environment." The idea is simply to outline these theories and make them understandable within their original context. Too many histories of ideas have simply traced the evolution of certain

35. I have used general histories of psychology with a more or less "internalistic" approach such as the Italian history of psychology by Lazzeroni (1972), a French history by Reuchlin (1957, 1980), a German history by Lück, Miller, and Rechtien, eds. (1984), American histories by Wertheimer (1970) and Sahakian (1975), and Dutch histories by Verbeek (1977) and Boon (1982). A more contextual approach toward this early period can be found in the German histories by Jaeger and Staeble (1978) and Lück et al. (1987), and the Dutch history by Bem (1985). I have also used general histories of sociology with a more or less internalistic approach such as a French history by Aron (2 vols., 1965–67, 1969–71), a German history by Jonas (4 vols., 1972–4), a British history by Abraham (1973), a Polish-American history by Szacki (1979), and a Dutch history by Rademaker and Petersma, eds. (1974). A more contextual approach toward this early period can be found in the American book by Coser (1971), whose approach has various elements in common with the present study. For a more general study on European psychosocial thought around the turn of the century, see the classical study by Hughes (1958/1974, among others). References to books on specific countries and themes can be found elsewhere in this study.

terms or approaches throughout the decades. They often tend to ignore that the meaning of these texts constantly changes with the changing contexts. I hope to demonstrate that their connotations can be deciphered correctly only by reconstructing something of their original settings, though there may of course be a difference of opinion on the question of which aspects are relevant and which are not.

In this regard, some may feel I have placed undue emphasis on current events – that is to say, on the years immediately preceding the writing of the text in question. My reason for this emphasis is that elaborate knowledge of these events is often *implied* in the texts but lost on today's reader. References to faraway events were often more explicitly elaborated, whereas topical references were considered self-evident and passed in silence. I hope to show that a contextual or even conjunctural approach to the history of ideas is imperative: Innovations can be assessed only when placed at a conjunction of lines of influence – past, present, and future.

After the optional sections on the early life and work of the author, and the regular sections on the social and intellectual context of his theories, we proceed to a section on the author's life and work at the time he wrote his monograph on the crowd, and then to the contents of that book or other works that were immediately related. I recapitulate them as much as space permits, in their original form and with as little commentary as possible on their scientific merits with hindsight. It is not my purpose to weigh the true value of their respective contributions to a "definitive" theory of crowd behavior, but to try and understand what their ideas meant at the time. This section on their crowd theories is the center of each chapter: There is a before and an after, the background and the impact of the book.

At this point in the chapter there may also be a section with a brief excursion into separate but related matters. This particularly holds for the first half of this study. The Taine chapter contains a brief section on his friend and colleague Boutmy, who founded the related field of political psychology. The chapter on Sighele contains a brief section on his contemporary and compatriot Rossi, who also wrote a long series of books on collective psychology and even founded the first journal of the field. And the Fournial chapter contains a brief section on the priority debate between Sighele, Tarde, and Le Bon, from which he was conspicuously absent, for reasons we will discuss later. Because this debate is not completely irrelevant (as some suggest), I have endeavored to reconstruct the debate from both public and private letters exchanged between these founding fathers. Some of this correspondence turned up in the archives of their heirs.

Finally, each chapter ends with a section on the main author's later life and work (except for Taine and Tarde, who did not survive the publication of their relevant works for a very long time), as well as on the significance and impact of his crowd theory either in science, ideology, or both. Taine became one of the most influential intellectuals of his generation. Sighele and many of the related Italian authors were active or became active in politics themselves. Le Bon inspired many of the major French and foreign political figures of the twentieth century. And Tarde should be considered one of the true inventors of social psychology and the interactionist approach to collective behavior. Yet, even in these last sections of each chapter, I hesitate to give a global value judgment on their contribution. That is not the purpose of this study.

According to Lorenz (1987), every attempt at historiography inevitably involves a certain "construction of the past," through selection and interpretation of relevant facts and aspects. Similarly, every discipline of the humanities is involved in the gradual construction of a certain view of psychosocial reality congenial to our culture. What I have tried to do, then, is to provide a fair reconstruction of how early modern crowd theories evolved.

1

The revolutionary mob: Taine, psychohistory, and regression

Menthon–Saint Bernard is a small town on the banks of Lac d'Annecy in France, some forty kilometers south of Geneva. It was in this quiet atmosphere that the noted historian Taine chose to write his magnum opus on the French Revolution and its *journées révolutionnaires:* the Réveillon riots, the storming of the Bastille, the marches to Versailles and Vincennes, the Champ de Mars demonstration, the invasion of the Tuileries, the overthrow of the monarchy, the September massacres, the expulsion of the Girondin deputies, the September insurrection, the overthrow of Robespierre, the popular riots of Germinal and Prairial, and the royalist uprising of the Vendémiaire.[1]

Some ten years before Guy de Maupassant wrote the lines quoted in the introduction to this study, Hippolyte Taine, who shared his pessimistic outlook,[2] painted the "spontaneous anarchy" that broke out after the taking of the Bastille. "However bad a particular government may be, there is something still worse, and that is the suppression of all government. For . . . it serves society as the brain serves the living being," he said.

> In such a state of things white men are hardly worth more than black ones; for, not only is the band, whose aim is violence, composed of those who are most destitute, most wildly enthusiastic, and most inclined to destructiveness and to license, but also, as this band tumultuously carries out its violent action, each

1. Rudé, *The Crowd in History*, p. 94.
2. In *Distorting Mirrors*, Barrows suggests that De Maupassant and Taine may both have died from syphilis (p. 82, n. 30).

individual the most brutal, the most irrational, the most corrupt, descends lower than himself, even to the darkness, the madness, and the savagery of the dregs of society. In fact, a man who at the interchange of blows, would resist the excitement of murder, and not use his strength like a savage, must be familiar with arms, accustomed to danger, cool-blooded, alive to the sentiment of honour, and, above all, sensitive to that stern military code which, to the imagination of the soldier, ever holds out to him the provost's gibbet to which he is sure to rise, should he strike one blow too many. All these restraints, inward as well as outward, are wanting to the man who plunges into insurrection.

He is a novice in the acts of violence which he carries out. He has no fear of the law, because he abolishes it. The action begun carries him further than he intended to go. His anger is exasperated by peril and resistance. He catches the fever from contact with those who are fevered, and follows robbers who have become his comrades. Add to this the clamours, the drunkenness, the spectacle of destruction, the nervous tremor of the body strained beyond its powers of endurance, and we can comprehend how, from the peasant, the labourer, and the bourgeois, pacified and tamed by an old civilisation, we see all of a sudden spring forth the barbarian, and, still worse, the primitive animal, the grinning, sanguinary, wanton baboon, who chuckles while he slays, and gambols over the ruin he has accomplished. Such is the actual government to which France is given up. . . .[3]

Taine's dim view of the events on that epoch-making July 14 was published a few years before *Quatorze Juillet* was proclaimed *the* national holiday. Yet his view fit in well with the dominant ideas about mob behavior. It corresponded closely to descriptions by French novelists like Dumas, Flaubert, Hugo, and British novelists like Brontë, Dickens, Eliot, Scott: most of which he had probably read.[4] If their fictional prose was partly based on documentary evidence, however, his own "scientific" account was in turn partly based on literary license.

In this chapter, like subsequent ones, I try to answer the questions: Why this approach, why this subject, what did it mean, and how did it contribute to crowd psychology and related disciplines? Or more spe-

3. *The Origins of Contemporary France* (abr. ed.), pp. 122–4.
4. In view of Taine's own studies on French and English literature.

cifically: What was Taine's approach to the study of human behavior, and how did it develop? What caused his sudden interest in the French Revolution, and what formed his particular viewpoint? What were his methodological and theoretical concerns? How did he describe the revolutionary mob, and did this description correspond to the known historical "facts"? And finally: How did it affect early psychology, sociology, and political science?

Taine as a critic and psychologist

Although Taine's life and work covered a wide range of disciplines, it largely centered on historical and cultural psychology. It may be significant that he was born in 1828 – just before the end of the Restoration, and in the French Ardennes – close to the cultural frontier between the Latin Catholic and Germanic Protestant worlds. He came from a middle class professional family. His father, a lawyer, died when he was still a boy. His mother took him to Paris to attend the prestigious Collège Bourbon, and then the École Normale.

He gained his *licence-ès-lettres* (preliminary degree) in the year of the revolution of 1848, but failed his *agrégation* (advanced degree) in philosophy in the year of Louis Napoleon's coup d'état of 1851. The reason is said to have been that he spoke out too strongly against the prevailing spiritualist and eclectic philosophy, and in favor of a new materialist and empirical philosophy. After some hesitation, he changed his topic and gained his doctoral degree two years later.

Under the Second Empire, Taine kept politics at a distance. He partly lived on private means, traveled widely throughout Europe, studied languages, and published essays on foreign countries and peoples. But he also taught occasionally, and made a name for himself with a number of studies on Classical, French, British, and German writers and philosophers. In the process, he developed a highly personal approach, trying to penetrate the minds of these authors and the mentality of the epochs and societies they lived in.

In order to deepen his understanding of psychology, he turned to contemporary science. Charlton, in his study *Positivist Thought in France during the Second Empire,* calls him a typical representative of scientism, and therefore ranks him (along with Comte) among "the false friends of positivist philosophy" (p. 2). During the first half of the nineteenth century, France had contributed greatly to new developments in the natural sciences: in physics (Ampère), chemistry (Gay-

A schoolbook image of a lecture at the École de Médecine in Taine's day. (From G. Bruno, *Le tour de France par deux enfants*. Paris: Bélin, 1877/1912, p. 279.)

Lussac), biology (Lamarck), physiology (Magendie), and more. Under the early Second Empire, Bernard's application of the experimental method to medicine received widespread attention. There was a general call for a similar "positive" science of man and society, that is: one based on "hard" facts. This trend was stimulated by new developments in historiography and ethnography.

The central question was how material conditions affected the physiological makeup of peoples, and how these changes resulted in the evolution or degeneration of stock. The study of heredity came to the fore. From the early 1850s on, Taine had attended lectures on botany, zoology, anatomy, and physiology at the Sorbonne faculty, and had taken notes on psychopathology at the Salpêtrière clinic. In those days, he already felt that "true psychology is a magnificent science on which the philosophy of history is based."[5] In the process of studying foreign societies and the past, he therefore tried to develop some kind of historical and cultural psychology.

The best known statement of his ideas in these fields can be found in

5. Giraud, *Essai sur Taine*, p. 36.

the famous introduction to his *Histoire de la Littérature Anglaise* (1863). He claimed that in order to understand a writer we should not limit ourselves to the text of his writings or even the exterior aspects of his life, but should try to reconstruct his interior life as well: that is, his psychology, and that of his contemporaries and compatriots. Individual and collective passions might well be studied in the same scientific way as chemical reactions, he claimed.

"Vice and virtue are products, like vitriol and sugar; and every complex phenomenon has its springs from a more simple phenomenon on which it hangs."[6] The following year he reconfirmed to his close friend De Witt (son-in-law of the Orleanist statesman Guizot): "Now my philosophical idea is this: that all the feelings, ideas, and conditions of the human soul are products, each with its causes and its laws, and that the whole future of History consists in the research of those causes and laws. The assimilation of historical and psychological research to physiological and chemical research – such is my object and master idea."[7] He thus favored the study of individual and collective mentalities. "Three different sources contribute to produce this elementary moral state," Taine said: *la race, le milieu, et le moment* (race, environment, and epoch). "What we call the race are the innate and hereditary dispositions which man brings with him to the light" (p. 10). But man is not alone in the world; nature surrounds him, and his fellow-men surround him; accidental and secondary tendencies come to place themselves on his primitive tendencies, and physical or social circumstances disturb or confirm the character committed to their charge" (p. 11). "Beside the permanent impulse and the given surroundings, there is the acquired momentum. When the national character and surrounding circumstances operate, it is not upon a tabula rasa, but on a ground on which marks are already impressed. According as one takes the ground at one moment or another, the imprint is different; and this is the cause that the total effect is different" (p. 12).[8]

A year later, Taine was made professor of art history at the École des Beaux Arts. But his psychological approach remained somewhat sketchy, and he felt a strong need to elaborate it more thoroughly along scientific lines. He set out to do so in his study *De l'Intelligence* (1870). It opposed philosophical speculation and the introspective

6. English ed., *Life and Letters,* Vol. I, Intr., sec. 3 (p. 6). 7. English ed., Vol. II, p. 255.
8. English ed., Vol. I, Intr., sec. 5 (pp. 10–12). Léger pointed out that Taine had already developed this famous formula a dozen years earlier. Also see Sternhell, *La Droite Révolutionnaire,* p. 156.

method, and advocated empirical research and experimental verification or case studies. The fact that he was well aware of some of the English and German psychological literature of the day also helped him identify the main themes of the prospective field. On the one hand, he urged a more systematic study of psychological functions: a program realized in part by his younger friend and colleague Ribot. After having summarized the state of the art in England and Germany, and after having written a major study on psychological heredity, Ribot published successive monographs on memory, will, personality, attention, sentiments, and other topics in psychology. On the other hand, Taine drew attention to unconscious phenomena in illusions and dreams, hallucinations, and hypnosis: a theme that was subsequently explored by the elder Charcot and the younger Janet. Taine said in a celebrated metaphor, that the human mind is a theater where several different pieces are being played at the same time, on different stages – only one of which is in the spotlight.[9] We will return to these psychological developments in the chapter on Le Bon. Freud, by the way, read Taine's book during a key phase of his intellectual development, and wrote to his friend Fliess that "it suited him well."[10]

For a number of reasons, Taine's book may therefore well be considered the starting point of scientific psychology in France. It was relatively successful, too: No less than twelve thousand copies were sold before the turn of the century.[11] After this study on cognition, Taine had intended to proceed with a sequel on volition. But he changed his mind and began to study the development of the French mentality instead. While doing so, however, he avidly followed new attempts to draw conclusions from evolutionary biology for human psychology and sociology. Thus, he was familiar with Spencer's *Principles of Psychology*, immediately read *Principles of Sociology* when it came out, and called the author "the most contemplative spirit [and] the greatest generalizer [*généralisateur*] of Europe" (letters to Gaston Paris and to Charles Ritter of July 1877). In his own *Last Essays*, too, he praised Spencer's thought as "new and fertile," because he had applied to the history of both nature and the mind the two key concepts that had been renovating the positive sciences for thirty years: that is to say Mayer and Joule's conservation of force and Darwin's natural selection (p. 118). Thus Taine's historiography became imbued with an evolutionist psychology and sociology built around such concepts as

9. Page 16 in the French edition. 10. In 1896. See Brandell, *Freud*, p. 75.
11. Giraud, *Essai sur Taine*, p. 174, n. 1.

struggle, progress, and regression. Darwin had used the concept of retrogression to indicate "backward development." Taine felt that a people that gave itself over to crowd excesses was sliding back on the ladder of civilization. This intuition was later formalized by Spencer's younger colleague Jackson and Taine's younger colleague Ribot as the mechanism of "dissolution." It was assigned a major role in all early crowd psychologies, ranging from those of Sergi and Sighele in Italy, to those of Lacassagne and Fournial – and most of all Le Bon – in France.

The Franco-Prussian War and the Commune revolt

Taine suddenly changed his mind because of dramatic events in 1870 and 1871. Although the Second Empire had gradually moved from a purely authoritarian to a somewhat more liberal phase, social and political unrest had continued to mount. In 1870 Emperor Napoleon III seized on a dispute over the succession to the Spanish throne as a pretext to declare war on Prussia. It was a grave miscalculation. The improvised French armies were no match for Bismarck's well-oiled war machine. Within six weeks, the emperor and 80,000 of his men were taken prisoner at Sedan. When the news reached Paris, crowds invaded the national assembly at the Palais Bourbon and urged Gambetta and other members of the opposition to lead them to the Hôtel de Ville, where a republic was proclaimed.

The Prussians continued their advance, however, and laid siege to the city. As a representative of the provisional government, Gambetta floated over the German lines in a balloon, established himself in Tours, and later in Bordeaux. He helped create a united front and mobilized resistance. But the situation in Paris continued to deteriorate. Hundreds of thousands of middle-class families fled to the provinces, and the lower classes emerged as a major force. Stirred with patriotic fervor, radicals called for a *levée-en-masse*. When the liberal conservative Thiers came to Paris to propose negotiations, for instance, a mob surrounded the Hôtel de Ville and demanded the continuation of the war. But the winter of 1870–71 approached, and the situation continued to deteriorate further. Cold, hunger, and disease took some 25,000 lives in the capital. After a counterattack failed, and after the total number of French deaths in the war had risen to some 135,000, the government was forced to yield to the facts. A last minute

Battle between loyalist and rebellious guardsmen in front of the Paris Hôtel de Ville on January 22, 1871, as seen by the *Illustrated London News*. [From G. Soria, *Grande histoire de la commune*, Vol. 1. Paris: Laffont, 1970, p. 285. Original in *Illustrated London News*, Vol. 58, pp. 176–7 (Bibl. de l'Arsenal).]

attempt by radical guardsmen to prevent an armistice failed, and it was finally signed in late January 1871.

Nationwide elections in early February reconfirmed the deep rift. They brought only very limited results for Bonapartists and Radical Republicans, who were held responsible for the disaster; modest results for the so-called Opportunists wavering between a conservative republic and a constitutional monarchy; and a monarchist landslide for both Orleanists and Legitimists. Thiers, who was acceptable to both the conservative republicans and the constitutional monarchists, was made the chief executive.

His government was forced to accept the Prussian conditions for a peace treaty: the cession of a large part of the rich Alsace-Lorraine region, and the promise of a punitive "indemnity" of five billion gold francs. But demonstrations and riots signaled continued opposition.

Proclamation of the Commune in front of the Paris Hôtel de Ville, on March 28, 1871. [From A. Castelot and A. Décaux, *Histoire de la France et des français au jour le jour*. Paris: Perrin, 1977. Original by Lorédan Larchey, *Mémorial des deux sièges de Paris*, Pl. 11 (Bibl. de St. Denis).]

When Thiers cut off their pay, Paris Guards concentrated their cannon in their stronghold, on the hill of Montmartre. And when he sent troops to seize them, soldiers fraternized with the crowd, and two of the generals were killed.

Thiers then moved his government back to Versailles as a precaution, and the Parisian radicals proclaimed autonomy. Many of their leaders had been influenced by the elitist ideas of the utopian socialists and felt they should lead the masses to change, have the capital set an example to the country, and complete previous revolutions. Thus in March 1871 they proceeded to elect a new city council or Commune, which undertook immediate reforms: a revival of the revolutionary calendar; free, compulsory lay education; freedom of assembly, association, and the press; popular election of all officials; abolition of conscription and the standing army. The liberal measures in the eco-

Radical guardsmen evacuating the Catholic Saint Sulpice Church on the night of May 12, 1871. (From G. Soria, *Grande histoire de la commune*, Vol. 3. Paris: Laffont, 1970, p. 272. Original by Lorédan Larchey in *Mémorial des deux sièges de Paris*, Pl. XXI.)

nomic sphere that followed later made these reforms a model for socialists of generations to come.[12]

The Versailles government reacted to this insubordination by arresting some Parisian members of parliament and by executing prisoners An abortive march of radicals on Versailles led to even more prisoners and executions. The Paris Commune thereupon seized the archbishop and several dozen priests as hostages. The stalemate lasted for six weeks, during which Versailles gradually assembled an army of 130,000 soldiers to crush 30,000 poorly trained and ill-equipped Paris guardsmen, and their popular support.

12. See, e.g., Marx's *The Civil War in France* (Marx and Engels, 1974–) and Lenin's *On the Paris Commune* (1961–8).

Execution of Communard prisoners. (From G. Soria, *Grande histoire de la commune,* Vol. 4. Paris: Laffont, 1970, p. 282. Original in *L'Illustration,* Vol. 57, p. 328.)

On May 21, 1871, they entered the city from the west, thereby opening the infamous *Semaine Sanglante*. Barricades sprang up throughout the city, but Haussmann's boulevards provided a clear range for heavy artillery. Commune prisoners were shot as they surrendered to the Versailles troops. Extremists in turn got hold of the hostages and killed them. Mobs lynched several dozen others, whereas *Les Pétroleuses* reportedly set fire to the Tuileries, the Louvre, the Palais Royal, the Palais de Justice, and the Hôtel de Ville as they were forced to retreat. On May 27, some of the last Communards died with their backs to the *Mur des Fédérés* on the Père Lachaise cemetery, whereas others fell on the Butte Chaumont.

It had been an unequal struggle. On the whole, the Versailles government lost 1,000 combatants, against the Paris Commune's 4,000. But at least 20,000 other Parisians were killed, many after they had been captured. Of the almost 40,000 others who were arrested, more than a quarter were sentenced to prison or deportation, and a third of these never returned.[13] Meanwhile a heavy tax was levied on the

13. Details in Rougerie, *Procès des Communards.*

remaining population of the capital, in order to finance the building of a Catholic *chapelle expiatoire* on the exact location where the mutiny had started. This monumental white Sacré Coeur cathedral, standing at the top of Montmartre, dominates the restive metropolis to this day.

The French elite was severely shocked by this new upsurge of the urban masses, which after each seemingly fatal blow, seemed to rise up again and again like a seven-headed hydra. A wide spectrum of politicians ranging from Thiers to Gambetta himself denounced the destructiveness of the masses and crowds. They were joined by most of the intelligentsia, both at home and in neighboring countries. Literary authors such as Dumas fils, Flaubert, the surviving Goncourt brother Edmond, and others expressed their disgust.[14] Scholarly authors tried to reconstruct the course of events and relate it to the latest discoveries of medical science. The Bibliothèque Nationale contains several hundred volumes on the subject, published during the 1870s alone.[15] They range from Laborde's improvised booklet *Les Hommes et les Actes de l'Insurrection de Paris devant la Psychologie Morbide* (the people and acts of insurrection in Paris before psychopathology; 1872), which significantly opened with a chapter on *folie collective*, to Du Camp's multivolume and semiofficial study *Les Convulsions de Paris* (published from 1878 on).[16]

Several of these authors were close friends of Taine, who shared their outrage. In his letters from these years, mostly to his wife and his mother, we can follow his reactions to the events from day to day. From the start, he emphasized the pathological nature of it all. His first reaction to the proclamation of the Commune, for instance, was that the new leader of the national guard was a "madman" (letters of March 19 and 24, 1871). Later, he opined that all the *fédérés* were "furious" (April 5) and "madmen" (April 20). At certain times, he stuck to the notion that they were mostly marginals and said there were maybe fifty thousand socialists, terrorists, "casse-cou, gens-sans-aveu, déclassés" of all sorts "at thirty pennies a day" (March 25). But at other moments, he felt that they were supported by a large part of the population, and that the whole of Paris had gone mad (March 26), possibly because of the long siege by the Prussians (March 20).

14. See, e.g., the ingenious (re)construction of the conversations they (may have) had in Baldick's *Dinner at Magny's*, particularly in the sixth chapter.
15. Rougerie, *Procès des Communards*.
16. See Cochart's article "Les Foules et la Commune."

In the end, however, he came to feel that the problem was even more serious than that: Something was thoroughly wrong with the national character of the French as a whole. When their daily routine was disturbed, the French people became easily disoriented and lost their minds (April 5). "I arrive at the following conclusions. Firstly, that the guiding principle of such minds must be a strong egoism. . . . My second conclusion is that in difficult questions, as those of Government, society, political constitution, the intelligence of the average Frenchman is at fault. His point of view is very limited; he is content with windy eloquence; he prides himself with his incompetence, and does not even realize the delicacy and the abstruseness of the question. And, in default of political intelligence, he has not the political instinct of the Englishman or of the Northerner in general" (April 8).[17] In brief, the vulnerability of the French to fits of collective madness and their other weaknesses did have strong political implications.

Meanwhile, Taine had left the Paris region for Tours. From there he traveled to Oxford, where he was to receive an honorary degree. It was at Oxford that he learned of the unraveling of the revolt. After his return, he set out to explore the political implications in a long article, which was published six months later in the liberal-conservative daily *Le Temps,* and in 1872 reprinted as a brochure.[18] This essay, *Du Suffrage Universel et de la Manière de Voter* (on universal suffrage and the manner of voting), contains a key to his political turn of mind at this point in time. France's experience with universal suffrage, he said, had been rather disappointing. Yet, it would probably be impossible to abolish it. It was therefore of the utmost importance to attenuate its impact by abolishing the direct election of popular representatives and having the voters choose electors instead, who would then select deputies. This would provide a better guarantee against excess. In an unpublished note he added that he would further elaborate these ideas in a major new study, upon which he was about to embark.[19]

In the first outline of this work, he was even more explicit. On the one hand, he said, universal suffrage is a system in which the people and the populace – that is to say mere numbers – dominate. On the other hand, the system of the *scrutin de liste* benefited only the candidates of the crowd and excluded refined and well-educated people. If

17. English ed., *Life and Letters,* Vol. III, p. 39.
18. Also included as an appendix in Mongardini, *Storia e Sociologia nell'Opera di H. Taine,* pp. 399–416.
19. Quoted by Giraud, *Essai sur Taine,* p. 90, n. 3.

the country stuck to these systems, he said, the riff-raff and the demagogues would end in dominating the prosperous and honest citizens, and heavily taxing them.[20] This could only lead to further disasters. It was therefore imperative to find out how these conceptions had come about, and what they had wrought so far. It was with these ideas in mind that he finally embarked on his major work: his multivolume study *Les Origines de la France Contemporaine*.

French historiography in transition

The lost war of 1870 and the abortive revolution of 1871 provoked a thorough reorientation of French thought. The entire subsequent period was dominated by an *examen de conscience*. Among other things, French intellectuals suddenly became aware of the superiority of German science, in both education and research.[21] This held not only for physics but for the humanities as well. One such exemplary discipline was closely related to Taine's original field: philology, the critical study of original literary and historical documents and materials. It was thought to be the basis for a new historical and social science, which should be introduced and adopted in France as well.

Of course a sense of history had slowly become more marked over the previous decades. Part of this awareness was a general European phenomenon, related to the fourfold technical, economic, social, and political revolutions as well as to the reality of rapid change and regional differences. But part of it was specifically French, too. No other country had been so unstable politically. For a full century, each generation had witnessed a radical change of regime in Paris: in 1789–94, 1802–4, 1814–15, 1830–1, 1848–51, and finally 1870–1. Why did the country not change gradually, as in Great Britain? Why did not the country grow stronger, as in Germany? The answers to these questions were sought in the study of the rise and fall of great civilizations in general, and the origins of contemporary France in particular.

Den Boer (1987) has reconstructed the gradual professionalization of French historiography, a process that of course began well before. But he says that after some preparatory work under the July monarchy, the Second Empire was the *haute époque* of the inventorization of materials and documents (p. 115). Monuments were classified and

20. Appendix to the third volume of the French edition, pp. 324 and 350, respectively.

21. See, e.g., Digeon, *La Crise Allemande de la Pensée Française*, and also Zeldin, *Intellect and Pride*, ch. 3.

restored, archives ordered and listed. The budgets of the Archives Nationales and of the Bibliothèque Nationale, for instance, rose by some two thirds in twenty years. The Third Republic, in turn, was the *haute époque* of education. The number of history teachers at *lycées et collèges* for boys rose from 108 to 618 during its first forty years, the number of history (and geography) chairs at literary faculties rose from 20 to 57. The number of students, *agrégés,* and dissertations rose accordingly.

The origin of modern historical studies in France is often traced back to the measures taken by Victor Duruy. He had originally been a history teacher at the École Polytechnique, then became a historical collaborator of Emperor Napoleon III himself, and finally an influential minister during the liberal phase of the Second Empire in the 1860s. He promoted scientific research and higher education, not only in the university faculties, but also at *grandes écoles* such as the École Normale Supérieure, which trained an elite of college teachers, and the École Pratique des Hautes Études, which trained an elite of advanced researchers. Among other things he founded the fourth (literary and historical) section of the École Pratique, which introduced the new philological methods on a wider scale.

During this time an entirely new generation of modern historians came to the fore, who were to preside over the publication boom of the Third Republic. Ernest Lavisse (1842–1922), for one, began as a ministerial assistant to Duruy, became a private tutor to the imperial prince, then spent a few years in Germany before reconciling himself with the republic and becoming an influential professor, first at the École Normale Supérieure, then at the Sorbonne. Gabriel Monod (1844–1912), his friend and colleague, also spent a few years in Germany before becoming a teacher at the new section of the École Pratique. Later he also became a professor at the École Normale and the prestigious Collège de France. Along with some other patrons, they started major source collections, book series (such as the Lavisse's *Histoire de France*), journals (such as Monod's *Revue Historique*), and other historical publications, and gradually built an entire network of professional historians.

Before they really began to take over in the course of the 1880s, however, historiography was still dominated by an intermediate generation, and most of all by three towering intellectual figures: Fustel de Coulanges, Taine, and Renan. They were all aware of the new philological methods, and tried to base their inquiries on a careful study of materials and documents. At the same time, however, they refused to content themselves with a simple reconstruction of historical events, or even to limit themselves to a particular time and place. They meant to study the rise and fall of civilizations and major nations in order to formulate general laws on the behavior of groups and individuals. In some respects this ambition faded with the next generation of specialized historians.

The youngest of these three authors was Numa Denis Fustel de Coulanges (1830–89). He had made a name for himself with a major study on the institutions, customs, and beliefs of Greece and Rome: *La Cité Antique* (1864). He later became a professor (and director) at the École Normale Supérieure, where he embarked on a similar venture: an *Histoire des Institutions Politiques de la France Ancienne* (1875 ff.). Both works meant to be based on the "objective examination" of "primary sources," but also meant to uncover "general laws." The introduction to the first volume of the latter work ended with the claim: "History is the science of social facts, that is to say sociology itself." Nor surprisingly, Fustel exerted a major influence on Durkheim and other founders of sociology.

The eldest of the three major historians of this generation was Ernest Renan (1823–92). He had studied for the priesthood but, after experiencing a crisis of faith, he had turned to the history of Western religion. His controversial *La Vie de Jésus* (1863) approached Christ as an extraordinary historical figure rather than the Son of God. After the lost war and the abortive revolt, Renan published a key pamphlet, *La Réforme intellectuelle et morale de la France,* which voiced doubts similar to Taine's on universal suffrage. He later continued his scientific work with major series, *Histoire des Origines du Christianisme* (1866–79) and *Histoire du Peuple d'Israel* (1887–93). They contributed greatly to a sociology and psychology of peoples and cultures. Renan became a professor at the prestigious Collège de France and one of the leading intellectuals of his time.

Taine was two years older than Fustel and five years younger than Renan. We have seen that he made a name for himself with the study of classical and modern literature, both in France and abroad. He also embraced the new philological principles, but in the aforementioned introduction to his *Histoire de la Littérature Anglaise* (1863) he claimed that, in order to be able to understand collectivities and individuals, one had to study "la race, le milieu, et le moment" as well. Thus, historiography should be based on psychology and vice versa. In the introduction of his study *De l'Intelligence* (1870), Taine claimed that "Between psychology thus conceived and history as it is now written, the relationship is very close. For history is applied psychology, psychology applied to more complex cases" (p. xi in the English translation). And in a letter written during these same years, he claimed that historiography should be based on psychological insight.[22]

When Taine began his study on the origins of contemporary France, this subject had of course been treated by many others before him. As a matter of fact, Gérard (1970), Godechot (1974), and others have shown that every regime had known several schools of thought on the French Revolution, its causes, and its effects. One such school was usually governmental in orientation and dominated official bodies and public education. It vindicated the current regime (and past regimes to which it declared allegiance), and in turn tended to denounce the previous regime (or other regimes it had replaced). Another school was often oppositional in orientation and largely depended on independent teachers and writers. It would make contrasting value judgments and often announce yet another regime change.

Under the Restoration, for instance, official historiography used to be integrist and legitimist in orientation. But at the same time, constitutional monarchists like Guizot, Thiers, Mignet, and others painted an image of the French Revolution that helped prepare the ground for the return of the House of Orleans in 1830. During the ultimate years of the July monarchy, by contrast, Left-leaning republicans like Blanc, Lamartine, and Michelet painted an image of the French Revolution that helped prepare the ground for the revolution of 1848. Each successive "History of the French Revolution" assigned a different role to the king and the aristocracy, the third estate, and "the people."

The renowned historian Jules Michelet, for one, created the central myth of the republican ideology in his monograph *Le Peuple* (1846). He also built his subsequent *Histoire de la Révolution Française* (1847–53) around the claim that the ordinary Frenchmen had embraced the ideals of the Enlightenment and spontaneously rose to shake off the yoke of obscurantism and absolutism by storming the Bastille prison. Many felt that this republican myth survived the Second Empire and had again resurged during the Commune revolt. Thus the liberal conservatives felt this notion of a heroic crowd overthrowing a re-

22. Appendix to the third volume of the French edition, pp. 303 and 316, respectively.

pressive regime was to be destroyed once and for all. A revised history of the French Revolution would have to make a careful study of the original documents that had become newly available. Continuing in the vein of Tocqueville's study on the *L'Ancien Régime et la Révolution,* it would have to show that long-term changes had been going on for decades, and that the revolutionary mobs had only made things worse.

The root of the problem, in Taine's opinion, was the idyllic view of human nature that had been developed by Rousseau and other Enlightenment philosophers. Their sterile utopianism had contributed heavily to the catastrophic turn of the French Revolution and subsequent events. Instead of learning from these experiences, however, the French had returned to make the same mistakes over and over again. Another reason was the rosy picture that republican historians such as Michelet had sketched of the "Great" Revolution. Although Taine was a fervent admirer of his style of writing history, he increasingly felt that its entire political perspective was dubious. For Michelet, the third estate had made the revolution and had led the people to victory, whereas the excesses had been committed by outsiders. For Taine, on the contrary, the third estate had lost control from the very start to the fourth estate and the *Lumpen,* the lowest scum of the big city. In his view, therefore, the mob had been the main actor.

This meant that crowd psychology was to play a key role in Taine's account. No good could come from mob rule, Taine felt, and instead of idealizing it, one should expose it in its full horror so that, instead of imitating these feats over and over again, people would put an end to them once and for all. Thus, it was of the utmost importance to finally write a "realistic" or "naturalistic" history of the Revolution, based on the latest discoveries of the sciences of man and society. If books had caused the Revolution and its endless repetition, then books could undo it.[23]

The origins of contemporary France in the French Revolution

In his preface to the first volume of *Les Origines de la France Contemporaine, L'Ancien Régime,* Taine once again discussed the central motive for his entire venture. On the one hand, the motive was France's recurring dissatisfaction with its political structure: "We have

23. The phrase is from McClelland, *The French Right,* p. 15.

demolished it thirteen times in eighty years." And on the other, it was the questionable solution to this problem: "The combined ignorance of ten millions is not the equivalent of one man's wisdom."[24] He then proceeded to dissect the ancient regime and on the basis of source material he set out to show that (1) the highly centralized social structure had been dominated by the king and nobility, who had, however, gradually ceased to render useful services in return for their privileges; (2) their customs and characters had increasingly revolved around a sterile *vie de salon;* (3) the classical spirit led to all kinds of simplifications about the nature of man and society; (4) the propagated doctrines aroused high expectations and revolutionary passions; and (5) the misery of the people made them highly receptive to these new ideas.

The key chapter (book V, ch. 3) was on the mental state of the people, and emphasized the role of crowd psychology. Taine claimed that the people had a "still rude brain" (p. 54), and "no instruction" (p. 55). Therefore "every idea, previous to taking root in their brain, must possess a legendary form, as absurd as it is simple" (p. 56). "Thus, among the dregs of society, foul and horrible romances are forged, in connection with famine and the Bastille, in which Louis XVI, the queen Marie Antoinette . . . the revenue farmers, the seigniors and ladies of high rank are portrayed as vampires and ghouls." The people "are like children," he said, with an "unconscious, apprehensive, popular imagination," "they have not the inward resources that render them capable of separating and discerning" (p. 57). In order to illustrate just to what extent Taine's historiography implied psychological and sociological notions throughout, I will quote extensively from other relevant fragments.

On the eve of the Revolution, the people had become restive, he said: At the moment of electing deputies "forty or fifty riots take place in one day" (p. 58). Thus "the furious animal destroys all, although wounding himself, driving and roaring against the obstacle that ought to be outflanked. This is owing to the absence of leaders and the absence of organization, a multitude simply being a herd" (p. 59). However,

> an insurrectionary multitude rejecting its natural leaders must elect or submit to others. It is like an army which, entering on a campaign, should depose its officers; the new grades are for the

24. *Origins*, pp. 3–4. The translator slipped here in making "quatre-vingts" twenty instead of (four times twenty, that is) eighty.

> boldest, most violent, most oppressed, for those who, putting themselves ahead, cry out "march" and thus form advanced bands. In 1789 the bands are ready; for, below the mass that suffers another suffers yet more . . . vagrants, every species of refractory spirit, victims of the law and of the police, mendicants, deformities, foul, filthy, haggard and savage, are engendered by the abuses of the system, and, upon each social ulcer they gather like vermin. (p. 60)

Taine's book caused a sensation: Never had anyone written about the events in this way. Translations were prepared; an American translation by John Durand (an abridged edition of which is quoted here) started coming out the very next year. Before the turn of the century, the book sold no less than 33,000 copies in France.[25] When it came out in late 1875 it fit perfectly into the continuing debate about the reasons for the Commune revolt and the nature of the new regime.

As a matter of fact, the national assembly had that very same year voted a new constitution, one that was nominally republican but did not really preclude a return to monarchy. The new conservative president MacMahon attempted such a legal coup d'état two years later, by appointing the Orleanist Duc de Broglie as a prime minister and trying to rig subsequent elections. But just like Taine, a large part of the educated classes had come to feel that such moves could only make things worse, and therefore reluctantly joined the "opportunist" (as opposed to the principled) republicans. After their candidate Grévy had been made president, however, it still took several years before key laws filled in the republican framework and stabilized the regime for the time being.

By that time, Taine had moved on, and started writing the volumes of *Les Origines* titled *La Révolution* (which expanded considerably in the process). His central theses were that (1) 1789 had marked, rather than a revolution, a dissolution of government and society; (2) it had resulted in spontaneous anarchy and mob rule, both in the streets and in the assemblies; (3) it had also alienated responsible citizens and paved the way for misfits and demagogues; (4) the Jacobin conquest of power was the result of systematic intimidation of anyone who dared to oppose them; and (5) in the end misery and despotism came to be worse under the revolutionary government than they had been under the

25. Giraud, *Essai sur Taine*, p. 174, n. 1.

ancient regime. This whole political process was primarily seen as the result of social and psychological mechanisms.

The first part, *Anarchy,* opened by describing how misery and hope had driven country folk to the cities and especially to the capital:

> All hover around Paris and are there engulfed as in a sewer, the unfortunate along with the criminals. . . . Already, before this final influx, the public sink is full to overflowing. . . . daily they grow bitter and excited. . . . you may anticipate the fury and the force with which they will storm any obstacle to which their attention may be directed. [p. 98] . . . The moment the Parliament of a large city refuses to register fiscal edicts it finds a riot at its service. . . . Clearly a new leaven has been infused among the ignorant and brutal masses, and the new ideas are producing their effect. [p. 99] . . .
>
> The starving, the ruffians and the patriots, all form one body, and henceforth misery, crime, and public spirit unite to provide an ever-ready insurrection for the agitators who desire to raise one. But the agitators are already in permanent session. The Palais Royal is an open-air club where, all day and even far into the night, one excites the other and urges on the crowd to blows. In this enclosure, protected by the privileges of the House of Orleans, the police dare not enter. Speech is free, and the public who avail themselves of this freedom seem purposely chosen to abuse it. The public and the place are adapted to each other. The Palais Royal, the center of prostitution, of play, of idleness and of pamphlets, attracts the whole of that uprooted population which floats about in a great city. [p. 104] . . .
>
> A new power has sprung up side by side with legal powers, a legislature of the highways and public squares – anonymous, irresponsible, without restraint, driven onward by coffee-house theories, by transports of the brain and the vehemence of mountebanks. . . . This is the dictatorship of a mob, and its proceedings, confirming to its nature, consist in acts of violence: wherever it finds resistance, it strikes. [pp. 107–8]

Children, women, and drunkards are particularly prominent.[26] At last,

> the fatal moment has arrived: it is no longer a government which falls that it may give way to another; it is all government which ceases to exist in order to make way for an intermittent des-

26. See pp. 83, 102, 142, 153, 233, among others.

> potism, for factions blindly impelled on by enthusiasm, credulity, misery and fear. Like a tame elephant suddenly become wild again, the populace throws off its ordinary driver [as it proceeds to storm the Bastille]. [p. 112]

Throughout these first chapters, "Spontaneous Anarchy," Taine repeatedly returned to the psychology of the mob. In the subsequent chapters, "The Constituent Assembly," he insisted on the psychology of mass meetings, on the role of demagogues and galleries in its chaotic proceedings. And he concluded the final chapters, "The Constitution Applied," with paragraphs on the general "state of mind," and the 'psychology of revolution' in general.

The second part of the volumes on the French Revolution was titled *The Jacobin Conquest*. A number of paragraphs in the opening chapter were devoted to the origins, the education, and "the psychology of the Jacobin." "The adepts in this theory come," Taine said, "from the two extremes of the lower stratum of the middle class and the upper stratum of the lower class." But the majority, "almost all of them well-established, steady-going, mature, married folks" must be left out.

> There remains a minority, a very small one, innovating and restless, consisting, on the one hand of people who are discontented with their calling or profession, because they are of secondary or subaltern rank in it . . . and, on the other hand, of men of unstable character, all who are uprooted by the immense upheaval of things. [pp. 166–7]. . . .
>
> In many of the large cities . . . there are two clubs in partnership, one, more or less respectable and parliamentary. . . and the other, practical and active, made up of bar-room politicians and club-haranguers. . . . The latter is a branch of the former, and, in urgent cases, supplies it with rioters. [p. 177] . . .
>
> Taking the whole of France, all of the Jacobins put together do not amount to three hundred thousand. This is a small number for the enslavement of six million of able-bodied men, and for installing in a country of twenty-six million inhabitants a more absolute despotism than that of an Asiatic sovereign. [p. 178] . . .
>
> [However], it is peculiar to the Jacobin to consider himself as the legitimate sovereign, and to treat his adversaries not as belligerents, but as criminals. [p. 204]

Thus, through successive stages, Jacobin intimidation goes from bad to worse, and results in the September massacres of 1792: "six days and five nights of uninterrupted butchery" (p. 235).

The third part of the volumes on the Revolution was *The Revolutionary Government*. Its central chapters discussed the "psychology of Jacobin leaders." "Of the three," Taine wrote, "Marat is the most monstrous; he borders on the lunatic, of which he displays the chief characteristics – furious exaltation, constant over-excitement, feverish restlessness, an inexhaustible propensity for scribbling, that mental automatism and tetanus of the will [!] under the constraint and rule of a fixed idea." Taine then proceeded to a more detailed diagnosis of the successive stages of the regime: "ambitious delirium," "mania for persecution," "confirmed nightmare," and "homicidal mania." "From first to last, he [Marat] was in the right line of the Revolution, lucid on account of his blindness, thanks to his crazy logic, thanks to the concordance of his personal malady with the public malady" (pp. 247–59).

Then Danton: "The graft on this plebeian seedling has not taken; in our modern garden this remains as in the ancient forest; its vigorous sap preserves its primitive raciness and produces none of the fine fruits of our civilization, a moral sense, honor, and conscience" (p. 264). It was he, therefore, who organized the revolutionary tribunals: "It towers before him, this sinister machine, with its vast wheel and iron cogs grinding all France, their multiplied teeth pressing out each individual life, its steel blade constantly rising and falling, as it plays faster and faster" (p. 268).

Last but not least: mediocre Robespierre, the *cuistre*, "steady, hard-working, studious and fond of seclusion" (p. 276): "At the end of three years Robespierre has overtaken Marat, at the extreme point reached by Marat at the outset, and the theorist adopts the policy, the aim, the means, the work, and almost the vocabulary of the maniac: armed dictatorship of the urban mob, systematic maddening of the subsidized populace, war against the bourgeoisie, extermination of the rich, proscription of opposition writers, administrators and deputies" (p. 285). Conclusion: "Suppress the Revolution, and Marat would have probably ended his days in an asylum. Danton might possibly have become a legal filibuster, a Mandrin or bravo under certain circumstances, and finally throttled or hung. Robespierre, on the contrary, might have continued as he began, a busy, hard-working lawyer of good standing" (p. 274).

The "naturalist" crowd psychology implicit in many of Taine's descriptions of the events during the *journées révolutionnaires* of the French Revolution was based on three elements. First, in such crowd events the wild forces of nature were released. Thus he had a clear predilection for comparing participants in these collective protests to more primitive forms of life such as animals and plants. Second, this release was depicted as the result of a step backward on the ladder of civilization. Thus he emphasized the eruption of violent emotions and the loss of rational control. Third, this regression apparently occurred more easily (but not exclusively) in people who he believed to possess less self-control anyway. Thus he highlighted the role of drunkards and criminals, but also of women and children. Barrows (1981) has given an excellent analysis of the persistence of such contemporary "metaphors of fear."

Taine's devastating volumes on the Great Revolution were published between 1878 and 1884, when the republic was gradually being consolidated, when the radicals returned to prominence and the socialists became active again. There was a resurgence of workers' agitation: of primitive unions, wild strikes, and violent outbursts – which prefigured some of the unrest of the nineties. Taine felt his books should be a clear warning against further revolutionary experiments, which would inevitably lead to new disasters. The volumes appeared in good time, too, to feed into the debate on the First Centenary of the Revolution and to dampen some of the republican euphoria on that occasion.

Taine did not live long enough to complete the entire work as projected, but he did finish most of the final volumes, titled *Le Régime Moderne*. The first parts of these were entirely devoted to the Empire and Napoleon, because "it is he who has made modern France." "Evidently he is not a Frenchman nor a man of the eighteenth century; he belongs to another race and another epoch" (p. 300). He is of the same flesh and blood as the princes that Machiavelli wrote about: "just at a time when the energy and ambition, the vigorous and free sap of the Middle Ages begins to run down and then to dry up in the shrivelled trunk, a small detached branch takes root in an island, not less Italian but almost as barbarous": Corsica (p. 302).

In his youth, Napoleon discussed some of the ideas of the Enlightenment "but these borrowed clothes, which incommodate him, do not fit him" (p. 305). So he goes on to become "a veritable *condottière,* that is to say leader of a band, more and more independent, pretending to submit under the pretext of the public good, looking out solely for his own interest" (p. 309). This exceptional plant, Taine said, encountered ideal conditions: "The soil of France and of Europe . . . broken up by revolutionary tempests, is more favorable to its roots than the worn-out fields of the Middle Ages . . . nothing checks its growth" (p. 315).

The main reason for Napoleon's success is, that he proves "himself as great a psychologist as he is an accomplished strategist. In fact, no one has surpassed him in the art of defining the various states and impulses of one or of many minds. . . . No faculty is more precious for a political engineer; for the forces he acts upon are never other than human passions" (pp. 323–4). But he exaggerates, and the main reason for his final failure is, that he is also an artist with an excessive imagination, according to Taine. Next to Dante and Michelangelo, Napoleon is "one of the three sovereign minds of the Italian Renaissance. Only, while the first two operated on paper and marble, the latter operates on the living being" (p. 336). It is he who built the modern French state.

In the last volumes of *Les Origines,* Taine went on to analyze the various aspects of the French state, and its relations to other institutions such as church and school. "In the establishment of an educational corps," Napoleon had said, "my aim is to secure the means for directing political and moral opinions." In order to do this, he even suggested keeping "notes on each child after the age of nine years" (p. 411). Thus, Taine felt, the centralized state had become excessively powerful and tried to subordinate all independent minds and bodies. In spite of all the changes that had taken place, a similar state of affairs had persisted under subsequent regimes. This had thoroughly influenced the French mentality.

Taine completed his naturalist portrait of Napoleon in 1886, just about the time when a new caesarist movement around General Boulanger started its rapid rise. As always, he published key psychological fragments in the form of articles. But the last volumes of *Les Origines* came out as late as 1891 and 1893. Taine died in the latter year. A few months earlier, he had reconfirmed to his biographer: "for forty years, I have only practiced applied or pure psychology."[27] Thus, his work on the French Revolution may best be seen as a precocious attempt at psychohistory or political psychology. In order to illustrate this, we will deviate from our self-imposed format here and devote a separate section to a venture in which Taine was closely involved: the founding of the École Libre des Sciences Politiques.

27. Giraud, *Essai sur Taine*, pp. xii, 1.

Boutmy and the origins of political psychology

Apart from being a theoretical venture, *Les Origines de la France Contemporaine* also served a more practical purpose. One of the main reasons underlying the governmental instability and revolutionary inclinations of France, Taine said in his outline, was "the insufficiency of the moral sciences," commonly associated with political science at the time. Their insufficiency sprang from the fact that they were not based on the empirical study of man and society, and on psychology in particular.[28] Thus Taine was involved in efforts to establish a school of political science, in which such studies were to be undertaken. In fact, as he later wrote to the imperial prince, "my book is but one among the documents that this institution will produce, a memoir to be consulted by coming statesmen."[29]

The central figure in this undertaking was Taine's close friend and colleague Émile Boutmy (1835–1906). He had been named after, and sponsored by, a business relation of his father's, liberal conservative industrialist and newspaper tycoon Émile de Girardin. This had helped Boutmy to become first a newspaper commentator, and then a professor in the history of civilization at the École Centrale d'Architecture.[30] In this period Boutmy came to know Taine, who was his colleague at the École des Beaux Arts; for some time they even worked on similar projects. After the dual catastrophe of the lost war and the abortive revolution, they both felt that new ways should be found to train a more competent body of diplomats and civil servants.

"It was the University of Berlin that triumphed at Sadowa," Boutmy wrote to a friend, and Taine (and Renan) felt the same. If Germany had won the war through its superior universities, they reasoned, then maybe France could redress the situation by improving its higher education as well.[31] If the established universities showed themselves too rigid to adapt, "free schools" might be set up by private initiative. In order "to enlighten the middle classes, to free them from the rule of a frivolous journalism" as Boutmy put it, and "to give the people back their heads" ("refaire une tête au peuple"). So Taine, Boutmy, and other liberal conservatives decided to raise a large sum, mostly from

28. Appendix to the third volume of the French edition, pp. 305 and 309.
29. English ed., Vol. III, p. 189.
30. On Boutmy, See Favre's article in Besnard, ed. (1981).
31. On school reform, see Elwitt, *The Making of the French Third Republic*, ch. 5.

the Orleanist financial and business elite, in order to set up an École Libre des Sciences Politiques. When the school continued to run a considerable deficit a few years after its founding, a duchess close to the pretender even came up with as much as a million francs (rewarded with a professorship for her son).

What should the curriculum of the new school be in order to realize these goals? Boutmy first made an inventory of the political science courses given within the framework of the established institutions and their shortcomings, while Taine made an outline for an alternative curriculum based on the new and "positive" sciences of man and society. The latter also promoted the project in the prestigious *Journal des Débats* by claiming that "science engenders prudence, and careful study diminishes the number of revolutionaries by diminishing the number of theoreticians."[32]

An early version of Taine's outline began with the study of physics and biology, but this had to be dropped for practical reasons. The final version began with the study of the main states, because their role in the world "is predetermined by the nature of their soil and climate, by the character of the race or races that compose them, by the religion that they have adopted."[33] It then proceeded with the study of resources, economy, and finance; international, constitutional, and comparative law; and the organization of the government, the military, and society. There were to be ten different series of twenty lectures each, spread out over two years. The school started with only six professors and ninety students but grew to include forty professors and some three hundred fifty students by the turn of the century. At that time more than 90 percent of civil servants passing the competition for top positions in four main administrative bodies were graduates of the École.[34]

Many of the liberal conservative backers of the school were Protestant Anglophiles. Boutmy, who had become its director, focused mainly on the comparison of law, society, and character in France, the United Kingdom, and the United States. His approach was heavily influenced by Taine's. This is particularly evident in his articles on political behavior in the Anglo-American world, which resulted in two

32. See Rosenbauer, *L'École Libres des Sciences Politiques*, and Osborne, *The Recruitment of the Administrative Elite in the Third French Republic* (esp. pp. 60–65).
33. Taine, L'École Libre des Sciences Politiques, *Journal des Débats*, Nov. 10, 1872. Reprinted in *Derniers Essays*, pp. 77–98 (my translation).
34. The four bodies were the Council of State, the Inspection des Finances, the Cours des Comptes, and the diplomatic corps (Osborne, p. 95).

books: *Essai d'une Psychologie Politique du Peuple Anglais au XIXe Siècle* (1901), and *Éléments d'une Psychologie Politique du Peuple Américain* (1902).

In the book on the English, for example, he began with geography, climate, and natural resources, and linked them to such psychological characteristics as initiative, foresight, and self-control, but also phlegm, introversion, and pragmatism. He then proceeded to show how individualism, competition, and tolerance manifested themselves in economic, social, and political life. Only thereafter did he try to define the role of family, property, and class, of the gentry, farmers, and workers. His book on the Americans had some trouble following the same model, of course, in view of the diverse conditions and recent immigration into that country. Both books had a certain success in France, but met with considerable criticism abroad. His vision of British and American character, the critics said, was rather distorted. Yet, with these two books he was one of the first to introduce a new notion: that of a "political psychology."

Taine's influence on crowd psychology and related fields

Taine's seminal work *Les Origines* was first reprinted in a complete edition in 1899. In subsequent years, various studies were published on his life, work, correspondence, and influence. Giraud estimated that by the turn of the century, some two million people had read several of his forty books, and eleven million had read at least one – not counting translations.[35] His influence in France and abroad was overwhelming. With Renan, he thought through the basic premises of liberal ideology, distanced it from radicalism, and reconciled it with conservatism.

Traditional conservatism upheld God and king. Taine and Renan, however, turned away from integral Catholicism and legitimism. They did not base themselves on religion and a supernatural order, but on science and a natural order instead. Science proved, according to Taine, that men were born unequal, and that this inequality was largely inherited. Society should be governed by the best. If elites failed to perform their duties, they would be swept away. If they gave way to the masses, society would be governed by the worst. Thus it was crucial to have well-prepared elites, who could lead and educate the masses. In

35. Giraud, *Essai sur Taine*, p. 171.

his eyes it was doubtful, however, whether the French were ready for universal suffrage and a parliamentary regime. Much depended on the nature of the government. These ideas of Taine became the mainstay of liberal conservatism, although they inspired the extreme right as well.[36]

Taine's contributions to science itself were primarily in the field of political history. Barrès said that Taine, even more than Fustel de Coulanges, had made him grasp "the realities in the history of this country." Sorel said that Taine had revolutionized the history of the Revolution. And Nietzsche simply called Taine "the first living historian."[37]

Taine's inspiration was even more crucial, though, for certain branches of the early sciences of man and society. First, he contributed decisively to the emergence of crowd psychology. Lombroso said that Taine had "truly been [his] master." Lombroso and Taine were among the most-quoted authors in the first monograph on the crowd by Sighele. Taine was also among the most-quoted authors in the subsequent monographs by Fournial and particularly Le Bon. Le Bon's book on the psychology of revolutions also owed much to Taine (see Nye's introduction to the American edition in 1980). The social psychologist Tarde said that *Les Origines* contained "if not the synthesis, at least all the united elements of a good psychology of crowds and classes."[38]

Second, Taine contributed decisively to the emergence of crowd psychology's twin, elite sociology. Its founder Mosca acknowledged that he owed many of his ideas to Taine; his theory of the ruling class was inspired by the first volumes of *Les Origines*.[39] Pareto was more critical, but defended Taine against attacks. Max Weber called Taine the "sharpest mind" that France had produced since Comte.[40] And Boutmy's pupil Ostrogorski (to be elected representative of the Constitutional Democratic Party in the first Russian Duma) published a classic book *La Démocratie et l'Organisation des Partis Politiques* (1902–3), with a foreword by Bryce. The first book on party democracy, it was published well before Michels's *Zur Soziologie des Parteiwesens in der modernen Demokratie* (1911; on the sociology of the

36. On their contribution to nationalism in France, e.g., see Curtis 1959, Nolte 1970, Sternhell 1978, among others.
37. Mongardini, *Storia e Sociologia*, p.i. Nietzsche, *Jenseits von Gut und Böse, Studienausgabe*, Vol. III, p. 150.
38. Bélugou, *Enquête sur l'Oeuvre de Taine*, quoted by Mongardini, *Storia e Sociologia*, p. 255.
39. Livingston, in his introduction to Mosca's *The Ruling Class*, p. ix.
40. Mongardini, *Storia e Sociologia*, p. i.

party system within modern democracy) identified the "iron law" of oligarchization.

Third, Taine contributed decisively to the emergence of political psychology, political sociology, and political science in general, and more in particular to the analysis of the functioning of revolutionary movements and totalitarian governments. Present-day political psychologists, however, seem to be largely unaware of this.[41]

Yet, the overwhelming influence of *Les Origines* on political, social, and psychological thinking has its dark side, too. In retrospect, Taine was no exception to the rule that observers often tend to reduce the complex process of a political revolution to a much simpler image, which suits their political preconceptions. We have seen before that his revisionist historiography corresponded to well-defined preoccupations and goals. From the very start, his "objective" method was heavily biased. He relied on written material all right but tended to forget that it emanated primarily from the literate social class.

He also claimed that "The most trustworthy testimony is that of an eyewitness, especially when the witness is an honourable [!], attentive and intelligent man, writing on the spot, at the moment, and under the dictation of the facts themselves" (that is: not engaging in polemics).[42] It turned out, however, that Taine's preferred sources were mostly aristocrats and bourgeois gentlemen who felt immediately threatened by the Revolution and its radicalization. Furthermore, he aggravated or embellished their reports at will, in order to illustrate "general laws." And he often left out the reasons for popular excesses, thereby artificially emphasizing the irrationality of the masses and crowds.[43]

Of course radical republicans thoroughly resented the widespread influence of this antirevolutionary view of modern France. The left gradually mounted a comeback, first of all within the Paris Municipal Council, and decided on a counterattack on the historical front. Whereas monarchist monuments had long dominated the capital, the Council now decided to erect new statues: first honoring the Enlightenment philosophers Voltaire (1879), Rousseau (1882), and Diderot (1886), and later even a statue for the revolutionary hero Danton. Societies to

41. See, e.g., the historical chapters in the handbooks and introductions by Knutson 1973, Stone 1974, Streiffeler 1975, Roloff 1976, Moser 1979, and Long 1981.

42. *Origins*, p. 73.

43. He hardly mentioned the approach of Brunswick's armies, for example, as the immediate trigger to the September massacres.

commemorate the centenary of the French Revolution were formed from the early 1880s on, and later united into a national organization. A year after the publication of Taine's last volume on the Revolution, and in view of the approaching centenary, the Council also decided to free considerable funds for the creation of a special chair for the History of the French Revolution at the Sorbonne faculty, and for the publication of some major new studies. They were to show the positive role of "the representatives of the Paris people."

From 1885 on, this chair was occupied by the radical historian Alphonse Aulard (1849–1928), soon promoted to professor. Like Taine, he had begun as a student of literature, and a journalist. It was through a study of great orators such as Danton that he became involved in the study of the French Revolution. He began the publication of a vast collection of materials on the roles played by the Comité du Salut Public and the Club des Jacobins; studies on more or less respectable revolutionary leaders like Danton; and helped build up the newly founded journal *La Révolution Française*. His multivolume works with *Études et leçons* on the French Revolution and its political history made a major effort to refute Taine on all essential points. He even tried to make Taine's work a showcase of poor philological craftsmanship.

He proclaimed the "Ten Commandments" for the objective historian, and accused Taine of having sinned against every single one of them. Taine could not possibly have gone through all the available archives during the periods he spent in Paris, Aulard said, and it was easy to prove that he had just picked out the bits and pieces that suited him. In the end, Aulard even devoted two years of his Sorbonne lectures and a whole book to *Taine historien de la Révolution française* (1907), which tried to demonstrate that there were serious errors in almost every single section of his work. Even Taine's oft-quoted comparison of the bloodthirsty revolutionary ideology to an ancient crocodile cult was based on faulty documentation, Aulard said. And he concluded that the celebrated *Origines* were altogether "useless to history."[44]

Others (like Cochin) retorted by saying that the pot was calling the kettle black: that Aulard had also been occasionally blinded by his ideological fervor, and had committed serious mistakes himself. Still others decided to sacrifice some of Taine's reputation as a "factual"

44. Aulard, *Taine – Historien de la Révolution Française*, p. 330.

historian, only to rescue some of his reputation as an original thinker. One of them was Paul Lacombe, who had earlier criticized the new historians' one-sided emphasis on "*l'histoire événementielle,*" and who advocated a more social scientific approach. In his two books *La Psychologie des Individus et des Sociétés chez Taine historien des littératures* (1906) and *Taine historien et sociologue* (1909) he maintained that although Taine often got his facts wrong, he had seen certain tendencies right. Such claims, however, were frequently denounced again by subsequent critics.

Left-leaning historians like Albert Mathiez, George Lefèbvre, and Albert Soboul (all of whom came to occupy Alphonse Aulard's chair at the Sorbonne faculty) showed revolutionary leaders and masses in a more positive light. The debate on Taine's crowds continues to this very day, not only in France but also in the Anglo-Saxon world. Rudé (1959, 1964), for one, has made detailed studies of the composition and behavior of the Paris crowds during the *journées revolutionnaires* on the basis of arrest records and other original sources. He concluded that "Taine's 'mob' should be seen as a term of convenience, or rather as a frank symbol of prejudice, rather than as a verifiable historical phenomenon."[45] Yet others, such as Cobb, have since maintained that these studies give too rosy a picture of revolutionary excesses.[46] Thus the true nature of crowd behavior during the French Revolution remains a subject of controversy two centuries after the events have passed.

45. Rudé, *The Crowd in the French Revolution,* p. 239.
46. *Reactions to the French Revolution* among others. See Nye 1980, p. xlii.

2

The criminal crowd: Sighele, criminology, and semiresponsibility

Rome 1891. The same year in which the old and respected historian Taine began the publication of the last volumes of *Les Origines* in Paris, the young and unknown law graduate Sighele published the first monograph on the criminal crowd in Rome: *La Folla Delinquente*. Taine was cited often, along with spokesmen for a new school of criminological thought in Italy. He also cited Tarde and other authors, who had written separate paragraphs or pages on the mob. From these heterogeneous elements, Sighele tried to compile a general picture of the processes taking place in "criminal" crowds, and the judicial consequences this should have.

"What comes about in an *evolutionary* way in ordinary life occurs in a *revolutionary* way in the crowd," Sighele wrote (the emphasis is his).

> The disorganization of character that starts slowly under the influence of bad examples, or on the instigation of an already perverted companion, and that (after having made one fall into vice once and set out on a road on which one cannot halt anymore) extends ever further, until it has completely changed the individual and destroyed his character – all this occurs in the crowd in very few moments. Instead of the slow and gradual dissolution, which turns an honest man into an occasional criminal, and later even a habitual criminal, an instantaneous dissolution takes place in the crowd, turning an honest man into a

passional criminal. It is in this way, in my opinion, that a large part of the individuals present in a crowd, lapse into crimes.[1]

This explanation of collective behavior was only one among many provided, however. Those historians of science who have looked for a coherent psychological theory in Sighele's work have often been somewhat disappointed. The reason is that Sighele used an eclectic mix of elements that were worked out much more elegantly by other authors, such as Tarde and Le Bon in France. Many of the ideas were there, but the overall framework was different. Sighele was no social psychologist or psychopathologist, nor did he want to be one. Sighele was an early criminologist, and his work should be seen in that light. When placed in that perspective, his work is much easier to understand.

Actually, Sighele's small book is part of a longer series of studies on social influence and criminal complicity, and it fits into a very specific intellectual and social context. As in the case of Taine, we will deal successively with the following questions. What was the discipline in which Sighele worked, and what were the major paradigmatic debates of the day? What were the concrete events that helped inspire the study, and what were their implications? What was the content of the study, and of others to which it was related? We will then briefly consider a related author who played a significant role in early crowd psychology: Rossi. And finally we will delve somewhat further into Sighele's later life and work, and especially into his political activities – which have been subject to frequent misunderstandings.

Throughout the literature, there is a marked tendency to lump together the early French and Italian authors on crowd psychology into one "Latin" or "Roman" school. This tends to overlook the fact that these authors came from opposite ideological horizons. The French authors were mostly conservatives; the Italians, by contrast, were mostly radicals. The former looked upon the years 1870–71, for instance, as that of the last abortive social revolution: a national catastrophe that should never be repeated. The latter looked upon the same year as that of the provisional completion of national liberation: an occurrence that begged a follow-up in social reforms. The taking of Rome was the

1. Part I, ch. 3, sec. 3 (my translation). I adapted some of the punctuation in an attempt to make this rather complicated sentence somewhat more readable.

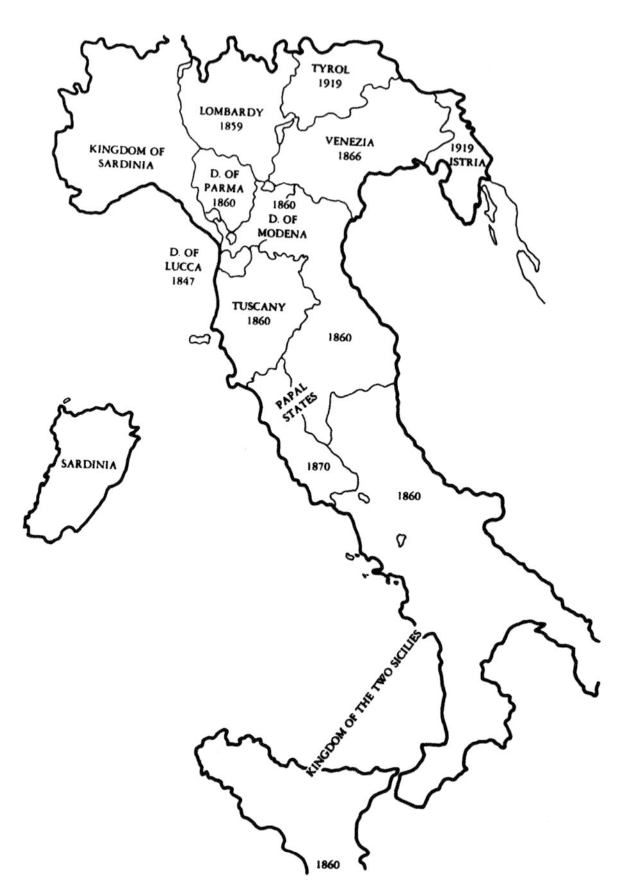

The unification of the Italian states 1847–1919. (From C. Tilly, L. Tilly, and R. Tilly, *The rebellious century, 1830–1930*. Cambridge, Mass.: Harvard University Press, 1975, p. 88. Copyright © the President and Fellows of Harvard College; reproduced by permission.)

crowning of the Risorgimento movement for the reunification of the country, and the road to further progress seemed to lie wide open.

This sentiment was reinforced by a major political shift in the ensuing years: In 1876 the "moderate" Left came to power around Agostino Depretis, and in 1887 the "pure" Left rose to power around Francesco Crispi.

> On the fall of the right, the Italian people, moved partly by the excessive burdens which had been imposed upon them, partly by what they had allowed themselves to believe of the wondrous

> power of progress, hailed with ecstasy the "new era," the "reforming government," the "righting of wrongs," the social "regeneration," the "healing" of all the ills from which they suffered – poverty, ignorance, injustice – and awaited with confidence the fulfillment of these vast promises,

according to one of the major historians of this period. "But sobriety and disillusionment soon followed."[2] The successive governments of the Left reduced female and child labor, abolished the unfair grist tax, introduced public health and social security measures, extended compulsory education and voting rights. But in spite of these measures, the scope of democracy remained fairly limited.

Some critics even claim that the Risorgimento itself had never been more than a half-hearted revolution from above. It had not been carried out by the masses, but by a small elite. The decisive battles had been fought by tiny bands and armies, directed by strong leaders: Mazzini, Garibaldi, Cavour, and the successive kings. With the approval of certain circles in Great Britain and France, these critics said, the northern states of Piedmont, Lombardy, and Liguria had driven the allies of Austria and Spain out of the southern states of Rome, Naples, and Sicily. In the name of the industrial cities and capitalism, they had subjugated the rural areas and feudalism. Between the midseventies and the mideighties, large-scale inquiries documented the miserable conditions persisting in the *Mezzogiorno* and the countryside, but few effective measures were taken to relieve their plight.

The outrage over the shortcomings of the Postrisorgimento regime was voiced most strongly by the left-wing opposition: the so-called *Estrema Sinistra*. It consisted of several currents. The "irredentists" demanded the further liberation of the *terre irredente,* that is to say of the northeastern territories, from Austrian rule. The republicans demanded abolition of the monarchy and further extension of voting rights. The radicals demanded thorough social reforms. In true Mazzinian and Garibaldian style, they were mostly active minorities trying to stir the 'silent majority' into action. They were partly successful. Gradually, nationalist and socialist mass movements arose, which challenged the newly established regime. In contrast to France, where none of the early major crowd psychologists was directly involved in politics, most of the early Italian crowd psychologists were.

2. Croce, *A History of Italy* (English translation), p. 71.

This means that we will have to deal with contemporary events in some more detail.

The emergence of positivist criminology in Italy

Within this general political context, a very peculiar intellectual climate developed. As in France, the major powerholders of the *ancien régime* had been closely linked to the church and religion. As in France too, a certain minority (including the Freemasons, for instance) had long cultivated secular ideas on man and society. Pannunzio said:

> In Italy the Comtean type of thinking had prevailed for centuries before Comte, through the works of Vico, Campanella, Machiavelli, Beccaria, and others, whose positivistic views were not unlike those of Saint-Simon and the direct founder of sociology. The ideas of "society as-a-whole," of advancement or "progress," of "equilibrium," or "cyclical movements" etc., abound in Italian writings (pp. 650–1). [But] concepts such as these, taking in as they do large units of time and space, do not conduce to the minute investigations necessary to building a body of scientific sociological theory and knowledge.

The Risorgimento changed all that. The Postrisorgimento elite turned to the universities and science in large numbers, and also embraced empirical and experimental research. Many claimed the country was backward, that it should try to catch up with the rest of Europe. Determinist, materialist, and evolutionist ideas were imported in large quantities from England, Germany, and France, and often vulgarized and combined in characteristically eclectic ways. Most of all Darwin, Spencer, Haeckel (and to a certain extent Lamarck) were felt to be the true heralds of a new science of man and society, and biology was widely considered to hold the key.

To many observers, the main social and psychological differences in the new Italy seemed closely related to physiological ones. Newly founded anthropological societies and journals claimed there were obvious morphological contrasts between the north and the south, and even between the upper and the lower classes.[3] Newly founded psychiatric societies and journals also claimed there were obvious phre-

3. These theses were elaborated by the positivist anthropologist Alfredo Niceforo in books on the Italians of North and South (1901) and the anthropology of the unpropertied classes (1910).

nological differences between normal and abnormal people. This view hardened when the belated translation of the works of Lavater and Gall on physiognomy and phrenology coincided with the new findings of Broca on cerebral localization.

One of the pioneers in this field was Giuseppe Sergi (Messina 1841 – Rome 1936), who had fought for Garibaldi at the age of nineteen, and who remained a radical thereafter. He had studied philosophy, literature, and geography. In 1873–74 (that is shortly after Taine, and at the same time as Wundt), he published a major work on experimental psychology: *Principî di Psicologia*. Five years later, it was followed by another work on physiological psychology: *Elementi di Psicologia*. He also developed stimulating ideas on the evolutionary origins of mental phenomena, the natural history of the sentiments, and the "stratification of character" (personality).[4]

After he had written a note to the education minister, pointing out the potential of the new sciences, he was made a professor in Bologna in 1880, and then in Rome in 1884. He first taught anthropology (where he introduced new methods of cranial measurement) and then psychology (where he established an early laboratory for experimental research in 1889, one of the very first outside Germany and the United States).[5] He soon became one of the central figures in the Italian sciences of man and society: He was involved in the founding of journals on scientific philosophy as well as education, psychology, anthropology, and sociology.[6]

Historians of psychology have often treated Sergi as merely another biologistic author, but his Spencerian theories were in fact closely related to those of his contemporaries Jackson and Ribot (see Chapter 4, on Le Bon) and contained similar psychodynamic intuitions. He claimed, among other things, that (1) psychological phenomena are just another function of the nervous system; (2) the nervous system itself is the outcome of a long history; (3) it consists of several strata formed during different stages of evolution; (4) the oldest and deepest (emotional) layers are well established in all human beings, whereas the more recent and superficial (rational) layers are well developed only in some; (5) emotional experiences are only a side effect of physiological reactions (this theory was developed simultaneously by James and Lange); and (6) the expression of strong emotions by some

4. Marhaba, *Lineamenti della Psicologia Italiana*.

5. See the list in Sahakian, *History of Psychology*, p. 139.

6. See Padovani, *La Stampa Periodica Italiana*, among others.

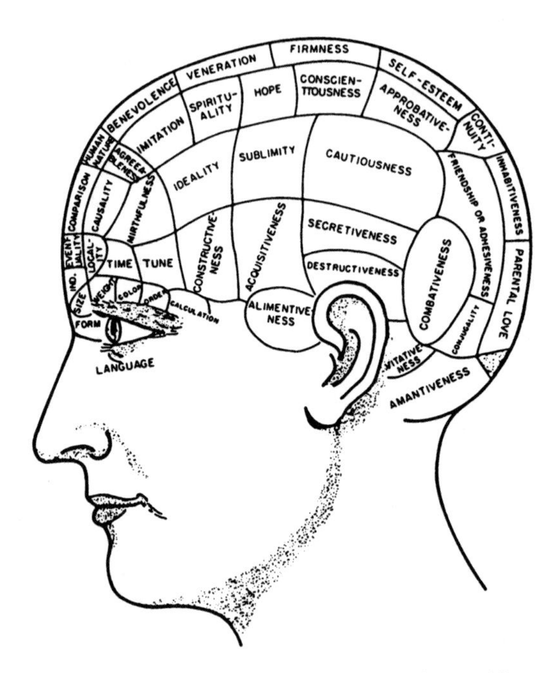

A phrenological chart, locating various mental faculties in specific parts of the skull. (From A. A. Roback and T. Kiernan, *Pictorial history of psychiatry and psychology*. London: Vision, 1969, p. 62.)

tends to elicit the same emotions in others. We will see later how these same themes recurred constantly in early crowd psychology.

Within the emerging sciences of individual and society, another discipline developed early on that attracted widespread attention: criminology. In various European countries, the growth of a modern judiciary and police corps, the improvement of administrative and statistical procedures had led to the gradual accumulation of comparable data on suicide, homicide, theft, and other offenses. This in turn stimulated a widening debate on the causes and prevention of crime.

Although Italy itself was relatively late in producing such data, the relative nature of legal arrangements was all the more obvious there. On the one hand, the unified nation had only recently been formed out of a number of smaller states, each with their own laws. Things that were considered a crime in one place might be legal in the next. Similarly, many of these states had gone through a century of upheavals: of alternating revolution and reaction, liberation and occupation. Things that were considered a crime at one moment might not be in the next, or vice versa. This situation had been highlighted well by an incident in the crucial year of 1859, when a deposed tyrant was recognized and lynched by a rebellious mob. The incident led to a heated debate, which was later picked up by people like Sighele.[7] He and most other early Italian crowd psychologists were primarily criminologists: They were interested in the question of how the group influenced the individual in such situations.

"The origins of criminology are generally dated from the late eighteenth century, when those imbued with a spirit of humanitarianism began questioning the cruelty, arbitrariness, and inefficiency of criminal justice and prison systems," according to Mannheim, a present-day authority in this field.

> From this period arose the so-called classical school of criminology, composed of such reformers as the Italian Marchese di Beccaria and the Englishmen Samuel Romilly, John Howard, and Jeremy Bentham, all of whom may be said to have sought penological and legal reform rather than criminological knowledge per se – that is, knowledge about crime and criminals. Their principal aims were to mitigate legal penalties and subject judges to the principle of *nulla poena sine lege* or "due process of law" and also to reduce the application of capital punishment and humanize penal institutions. In all this they were moderately successful, but in their desire to make criminal justice "just," they tried to construct rather abstract and artificial equations between crimes and penalties, thereby forgetting the personal characteristics and needs of the individual criminal. Moreover, the aim of punishment was seen as being primarily retribution, with deterrence occupying second place, and reformation lagging far behind.
>
> By the second half of the nineteenth century these deficiencies, together with the influential teachings of the French sociologist Auguste Comte, had prepared the ground for the positivist school, which sought to bring a scientific neutrality into crimi-

7. A former Parma army chief was lynched a few months after a referendum on affiliation with Piemonte. Political leaders such as Massimo D'Azeglio called for an "exemplary punishment" of the culprits (*Gazetta Piemontese*, Oct. 13, 1859). But juridical experts pleaded "limited responsibility" (*Gazetta dei Giuristi*, Oct. 21, 1859). The case was later discussed again in Zini's book *Storia d'Italia*, Vol. I (Milan, Guignoni, 1886), p. 442, and subsequently in Giurati's book *Gli Errori Giudiziarî* (Milan, Dumolard, 1893). The latter pointed it out to Sighele, who included it in the second edition of *La Folla Delinquente*.

nological studies. Instead of assuming a moral stand that focused on measuring the criminal's "guilt" and "responsibility" the positivists attempted a morally neutral and social interpretation of crime and its treatment. (Mannheim 1975, pp. 282–3)

The founder of this positivist school was Cesare Lombroso (Verona 1835 – Turin 1909). Today he is largely remembered for his reactionary theories of crime because of his biologistic approach, but he was neither clearly reactionary nor exclusively biologistic in his thinking.[8] His misconceptions become more understandable when placed in their original context, and even though they are largely discarded today they contributed decisively to the discussions from which modern criminology arose.

Lombroso studied medicine under Austrian rule (in Vienna, among other places), at a time when the influence of the phrenology of Gall and Spurzheim was still great. During the 1859 war of independence, he served as a physician in the Piedmont army. He began systematically measuring recruits from all over Italy, in order to make an inventory of regional variations in morphology. During the campaigns against banditry in the South, furthermore, he also became interested in the psychophysiological peculiarities of criminals. As the director of an insane asylum and a government consultant on mental health, he undertook studies on cretinism and pellagra, two widespread medical conditions believed to inhibit the normal development of both physical and mental traits and resulting in psychological disturbances often accompanied by violence.[9] Finally, he became a professor of legal medicine and public hygiene in Turin, later of clinical psychiatry, and later still in the new field of "criminal anthropology."

When performing an autopsy on a notorious brigand, he was struck by a cerebral anomaly that reminded him of apes. He thereupon under-

8. The equation biologistic = conservative may partly stem from a twentieth-century misreading of many nineteenth-century authors. We tend to take for granted their complete understanding of the Darwinian theory of evolution, which implies that "adaptation" of the organism to the environment takes very long. In this context, therefore, saying something is "inborn" almost equals saying it cannot easily be changed. At heart, however, many of these thinkers still really remained Lamarckians. They felt that with an amelioration of the environment, "stock" would soon improve, too. In many countries, furthermore, stressing evolutionary materialism was a way to oppose religious spiritualism. In this perspective, the adherence of many "progressive" thinkers of this period to biologistic ideas may look somewhat less incongruous.

9. *Encyclopaedia Britannica* 1975, Micropaedia Vol. V, p. 259, and Vol. VII, p. 840, respectively.

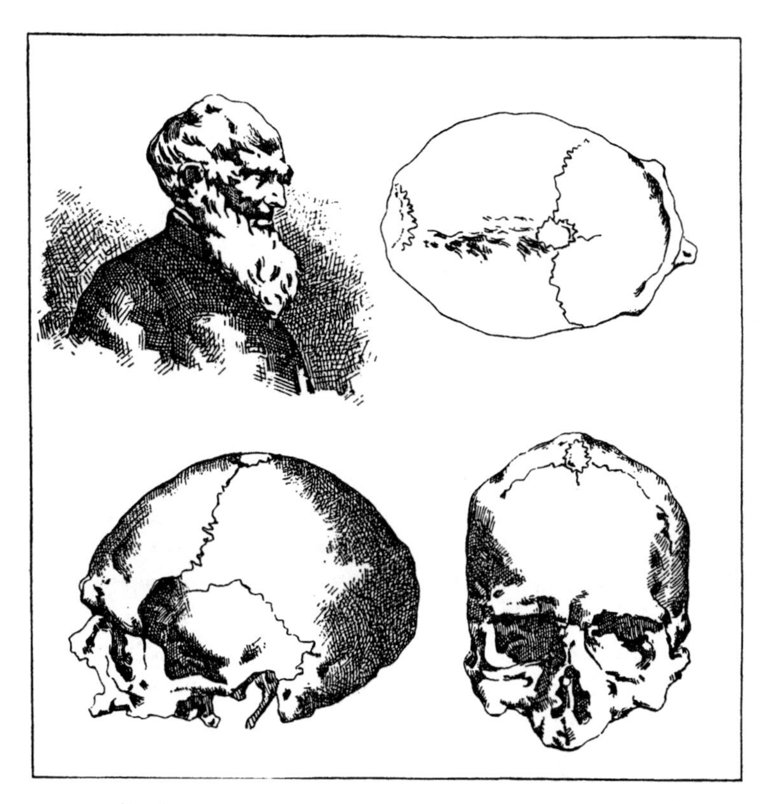

Portrait and skull of the bandit Gasparone (Fra Diavolo), as depicted in the Atlas added to the later editions of Lombroso's work on criminals. (From C. Lombroso, *L'uomo delinquente,* Vol. 5, 5th ed. Turin: Bocca, 1897. Atlante, Pl. XXX.)

took cranial and corporal measurements on a large number of delinquents, and came to the conclusion that many displayed "atavistic signs," such as deviations in head shape, the form of eyes, ears, nose, and mouth, the scale of the jaws and cheekbones, and the like. These also corresponded with psychological peculiarities, he said, such as acute perception and unrestrained passion, lack of morality and shame (as demonstrated by an absence of blushing).

After the publication of Darwin's *The Origin of Species,* the debate over evolution spread throughout Europe. In the late 1860s Haeckel first included man in this scheme, and in the early 1870s Darwin published *The Descent of Man.* He mentioned the possibility that "Injurious characters . . . tend to reappear through reversion, such as blackness in sheep; and with mankind some of the worst dispositions, which occasionally without any assignable cause make their appearance in families, may perhaps be reversions to a savage state, from which we are not removed by very many generations. This view seems indeed recognised in the common expression that such are the black sheep of the family."[10]

In his epoch-making 250-page work *L'Uomo Delinquente* (criminal man, 1876), Lombroso claimed that many delinquents were in fact predisposed to crime by their atavistic nature. The second edition included a much wider discussion of possible psychological and social factors, but the range was reduced again in subsequent editions and foreign translations – thus contributing heavily to his controversial fame.

Yet, new editions continued to expand: the fifth and final edition counted no fewer than five volumes and more than two thousand pages. In 1899 he added another volume primarily aimed at a foreign audience. In *Le Crime: Causes et Remèdes* Lombroso yielded to his critics somewhat, and assigned a larger role to environmental factors than before.[11]

The reactions of some foreign colleagues were mildly ironic. The famous anthropologist Westermarck later reported in his *Memories of My Life* how he had once visited a prison with him:

> Lombroso was obviously very popular amongst the prisoners, whom he treated as tenderly as a botanist does his precious plants. He joked with them and patted the cheeks of one or two of the women. "You are a murderer," he said to a bandit standing in his cell. "Yes, sir," was the cheerful answer. Last of all I was taken to the prison laboratory. Lombroso told me that one of his pupils had just made the interesting discovery that criminals often have projecting big toes. A prisoner who was present was asked to take off his shoes, but his big toes were quite normal. That was an exception. The test was made with another criminal, and with a similar result. By this time I was really anxious to beg leave to

10. Part I, Ch. V. Quoted from the second edition (1877, p. 137). Other related statements can be found in Darwin's notebooks, published by De Beer: "Our descent, then, is the origin of our evil passions!! – The Devil under form of Baboon is our grandfather!" (quoted in Sulloway, *Freud – Biologist of the Mind*, p. 242).
11. Other related books pointed out that genius is often related to madness, too (*Genio e Follia*, 1864), or that women also have their own criminal deviations (*La Donna Delinquente*, 1893).

show my own feet, which bore the distinctive mark of criminality; but my amiable guide would most likely have declared that they, too, were an exception. (p. 116)

Lombroso was a great admirer of Taine "the naturalist." He claimed, "Taine has truly been my master, the only master I have had besides Darwin."[12] Lombroso also wrote articles on crime in the French Revolution and its aftermath, which were clearly inspired by *Les Origines*. In 1890 he published a major book, *Il Delitto Politico e le Rivoluzioni* (political crimes and revolutions), with a younger collaborator. He maintained that a difference exists between political crimes and destructive revolts (that is to say, pathological phenomena) on the one hand and inevitable revolutions (that is, healthy phenomena) on the other.

The distinction was related not only to the level of support among the population but also to the presence of "philoneist" or "misoneist" tendencies – that is, the presence either of an exaggerated longing for, or an exaggerated fear of, social change among individuals or groups. Lombroso proposed to analyze the physical, biological, psychological, and sociological conditions underlying revolts and revolutions. Isolated political criminals, he said, were often predisposed to their deeds by inherent tendencies. A later book on "the anarchists" defended a similar thesis (which was of course related to ideological preoccupations).

Most reviews of his book on political crime were rather critical, including one by a young graduate in the *Archivio Giuridico*.[13] Much later, however, its author, Sighele, said that Lombroso's book had included "the entire theory of collective psychology" – albeit in fragmentary form.[14] Although this claim was clearly too generous, it is true that Lombroso exerted a powerful influence on Sighele. This influence was largely mediated, however, by Lombroso's closest associate and Sighele's immediate teacher: Ferri.

Enrico Ferri (San Benedetto Po, near Mantua, 1856 – Rome 1929) was much younger than Lombroso and held a much broader view. In 1877

12. In Bélugou, quoted by Mongardini, *Storia e Sociologia nell'Opera di H. Taine*, p. 257.
13. Vol. XLVI, Fasc. 6 (November 1890). Also published as a brochure.
14. In Bianchi et al., *L'Opera di Cesare Lombroso*, pp. 323–5. Also see Bulferetti, *Lombroso*, p. 331.

he had completed his law studies with a dissertation titled "Teorica della Imputabilità e Negazione del Libero Arbitrio" (theory of imputability and negation of the free will). Since human behavior turned out to be determined by so many factors, he said, penal law cannot remain based on the outdated notions of free will and moral responsibility. It should, on the contrary, be based on the right of society to defend itself, on an evaluation of the risk that the criminal poses to the public order, and therefore on the scientific study of the causes of crime.

This revolutionary thesis earned him a travel grant that took him to Paris, where he made an elaborate analysis of French crime statistics (the best in Europe at the time). This study and a subsequent one on homicide and suicide preceded some of the better known works by both Tarde and Durkheim on these subjects. Upon his return to Italy, he briefly stayed with Lombroso in Turin before being appointed professor in Bologna. His inaugural address, *I Nuovi Orizzonti del Diritto e della Procedura Penale* (the new horizons of law and of the penal procedure), covered 150 pages in its first edition of 1881. It was later renamed *Sociologia Criminale* and expanded to two volumes and almost twelve hundred pages in its fifth and final edition.

In it, Ferri identified five classes of criminals: "born" criminals as Lombroso had identified, insane, habitual, occasional, and passional criminals. He increasingly stressed the role of psychological and social factors in crime. In 1884 the second edition of *I Nuovi Orizzonti* even identified a new intermediate field between them.

> Between psychology which studies the individual and sociology which studies the society, I think there is room for a collective psychology, to study more or less defined groups. The phenomena of these groups are analogous, but not identical with those of the sociological body properly so called, according to as the union is more or less definite. Collective psychology has its field of observation in all unions, however occasional, such as the public street, the markets, workshops, theatres, meetings, assemblies, colleges, schools, barracks, prisons and so forth. Many practical applications of the data of collective psychology might be given. An example will be found in a future chapter, when I come to consider the psychology of the jury. (p. 94)

And on that subject he said: "if sometimes the jury can withstand the abuses of government, still too frequently it does not withstand its own passions, or the influence of the social class (the *bourgeoisie* in our own day) to which nearly all juries belong. . . . Besides, the same jury which will resist pressure from the Government does not resist popular pressure" (p. 182).[15]

From Bologna, Ferri went on to university chairs in Siena, Rome (1886), and finally the prestigious one in Pisa (1890); but he was dismissed a few years later on account of his political activities. Both Lombroso and Ferri were of modest northern origins and displayed radical sympathies, like many of their students. Only the third prominent founding father of the positivist school Garofalo (the author of the first book titled *Criminology* as such in 1885), who was of noble southern stock, distanced himself from these leanings. We will later see how these ideological backgrounds played a major role in the emergence of early crowd psychology in Italy.

The 1880s were the Golden Age of the positivist school in criminology. The group continually grew in size, and poured out a steady stream of articles and books, mostly published by the Bocca brothers in Turin. Many of the major titles were soon translated into French, German, English, or Spanish; their influence reached as far as Russia in the east, North and particularly South America in the west.

At the beginning of the decade, the "medical branch" of the positivist school founded its own scientific journal, the *Archivio di Psichiatria, Antropologia Criminale e Scienze Penali* (archives of psychiatry, criminal anthropology and penal sciences) "to serve the study of alienated and criminal man." Lombroso was one of the main figures, but many other national and international celebrities contributed at well. A few years later, Lacassagne (soon joined by Tarde and other French colleagues) founded a similar journal, which took a somewhat different line – to which we will return in the next chapter. Immediately after the end of the same decade, the "juridical branch" in Italy founded its own scientific journal as well: *La Scuola Positiva nella Giurisprudenza Civile e Penale* (the positive school in civil and penal jurisprudence "and in social life"). Here, Ferri was the main figure. By that time, however, the positivist school had become thoroughly enmeshed in violent polemics both at home and abroad.

In 1885 the school organized a First International Congress on Criminal Anthropology in Rome (parallel to a prison congress, which was

15. *Criminal Sociology* (English translation), p. 94 n. 1, p. 182.

held around that time in the same city). It was a "home match." The proceedings show that most foreign visitors were politely interested and only mildly critical. Some of the French, however, felt that the Italians were overprotecting the concept of the atavistic criminal. Although certain people might be predisposed to violence, they said, it had not been convincingly demonstrated that they shared common anatomical features.

When the Second International Congress of Criminal Anthropology was held in Paris in 1889, the roles were reversed, and the French launched a massive counterattack, pointing at certain methodological flaws in the documentary evidence. This spoiled relations between the Italian and the French schools for several years: precisely those years in which crowd psychology (and the related priority debate) emerged.[16]

During the late 1880s, the Italian school of positivist criminology was particularly keen on scientific recognition for at least one more reason. This was the current preparation of a new penal code in Italy. It was the ideal opportunity to try and get some of their theories put into practice. The Piedmontese code, which had been modeled on the French one, had been introduced in most of the Italian states, except Tuscany. It allowed for some special cases (insanity, blind rage, drunkenness), and the various projects for a new law also made criminal responsibility dependent on two conditions. First, one had to *know* what one was doing. Second, one had to be acting *freely* (and not under the influence of an "irresistible" force).[17] The positivist school felt that this did not go far enough, though, and that the final project remained too much within the limits of the "classical" tradition.

Thus Lombroso, Ferri, and their allies opposed the submission of the law to parliament, and the enactment by the government one year later. They stimulated their younger colleagues and students, instead, to write monographs on all kinds of extenuating or aggravating circumstances. One such study was written by Ferri's student Sighele, and dealt with the mysterious influence crowds seem to have on individuals.

16. See *La Scuola Positiva* 1892, pp. 423–4. Compare Nye 1984, pp. 103–9, etc. More details can be found in the next chapter.
17. See Paoli, *Esposizione Storica e Scientifica dei Lavori di Preparazione del Codice Penale Italiano*, p. 93.

The riots of the eighties, and the introduction of the semiresponsibility defense

The aforementioned abstract considerations form a major background of the first attempt to assemble and integrate early ideas on crowd psychology, but some concrete events played an equally important role. They were mostly related to the defense of people caught in riots.

It should be remarked at the outset that unrest was fairly limited during this decade, although the fear of anarchist violence remained. At least the unrest was limited in comparison to the preceding decade, when large-scale regional rebellions hampered national integration, and utopian insurrections undermined political stabilization. And it was equally limited in comparison to the succeeding decade, when proletarian revolts threatened the bourgeois regime and authoritarian reactions came close to eliminating parliamentary democracy altogether. The 1880s formed an intermediate period of relative calm. Although there was occasional agitation, it looked at the time as if it could and would gradually be absorbed. The state apparatus had grown larger and stronger, whereas the opposition movements were still relatively limited and weak.

As a matter of fact, most of the strikes and riots of the decade were related to the emergence of new forms of mobilization and organization among the lower classes. The emphasis was not so much on industrial workers, however, as on landless peasants – who were gravely affected by an agricultural depression, grain imports, price decreases, and the spread of capitalist farming. This held particularly for the Po river basin. According to the Marxist historian Procacci, "The labourers who worked in these places, who had often emigrated from neighbouring provinces, constituted a human and social aggregate to which there was no parallel in the agricultural proletariat of other European countries. . . . They were a newly formed social class, and in some ways their attitude was closer to that of workers and wage-earners than that of peasants."[18]

Some think this emphasis on class struggle and on the uniqueness of that situation is somewhat exaggerated, but there can be no doubt that this was the most restive region during these years. Tilly says, for instance: "A series of Po valley agricultural strikes came in geograph-

18. Procacci, *History of the Italian People*, pp. 343–4.

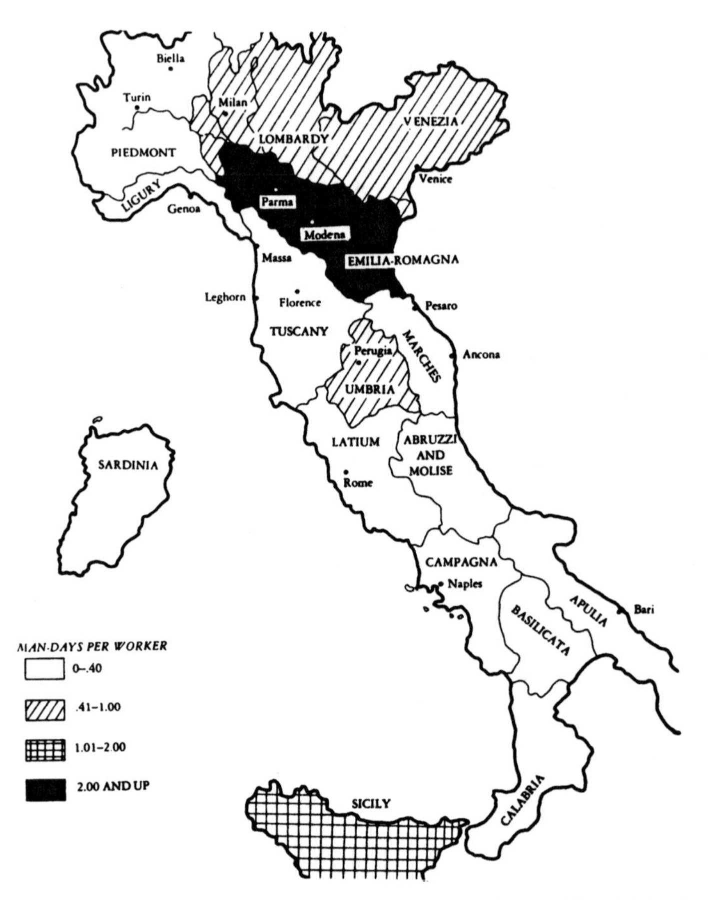

Strike propensity of agricultural workers, by Italian province, 1878–1903. (From C. Tilly, L. Tilly, and R. Tilly, *The rebellious century, 1830–1930*. Cambridge, Mass.: Harvard University Press, 1975, p. 160. Copyright © the President and Fellows of Harvard College; reproduced by permission.)

ically discrete waves throughout the 1880s. They were roughly related to organizing efforts by middle class reformers and to government repressive action. . . . In 1882 the Cremona and Parma regions were the scene of strikes; in 1883 it was the turn of Verona's hinterland. The Polesine . . . struck in 1884."[19]

In the spring of 1885, a strike was announced for the region of Mantua, but the authorities seized on a case of vandalism to crack down on the movement beforehand. After a few hundred arrests had

19. Tilly, *The Rebellious Century*, p. 145.

been made, the so-called *La Boje* rebellion broke out, quickly spreading to neighboring provinces such as the one around Milan, and resulting in a few thousand arrests. Most of the suspects were soon released, but twenty of the original "instigators" were kept in custody, and in early 1886 tried for "incitation to civil war."

One of their lawyers originated from the region: Enrico Ferri, whom we have already encountered as a founder of positivist criminology and an advocate of collective psychology. In order to be able to prepare their defense, he and his family moved for two full months to Venice, where the trial was held. He tried to demonstrate that the suspects had been understandably outraged by their miserable conditions and had mutually excited each other. And he invoked the "laws of human psychology" according to which "it is natural that the social unrest related to . . . the common longing for an improvement of their condition combined with a psychological fermentation, which is the result of the union of several wills, just like the mixture of several organic elements produces something which is called chemical fermentation."[20]

The court proved sympathetic to this argument, and on the first anniversary of their arrest, the suspects were freed. The judgment is also seen as the first recognition of the worker's right to form unions and to strike in Italy (although the new penal law expressly forbade "spreading class hatred"). Ferri grew immensely popular. Huge crowds acclaimed him in his native Mantua when he was elected Member of Parliament a month later. He got involved in setting up agricultural unions and a provincial federation with more than ten thousand members.[21] Other radical lawyers did the same in neighboring regions: Ferri's friend Bissolati in Cremona, Prampolini in Reggio Emilia, and so on. Thus, these events contributed decisively to the rise of a major mass movement.

During these years popular agitation was not limited to northern Italy, however, but affected southern Italy as well. In the very month of Ferri's election, major riots took place at the other end of the country: in Gravina di Puglia, in the heel of the Italian boot. On the holiday of Saint Michael's, the mayor of this town had at first allowed a religious service to take place but subsequently forbidden the secular activities

20. Ferri, *Difese Penali,* p. 44 (my translation).
21. Franchi, *Enrico Ferri,* p. 109; Ferri's "Autobiografia" in *La Folla* (1903).

"The Torrent of the Oppressed." Drawing by Galantara published in the Sunday edition of the socialist newspaper *Avanti*. (From B. Craxi (ed.), *Le immagini del socialismo 1892–1983*. Almanacco Socialista, PSI. Milan: Tip. Fiorin, 1984, p. 230. This drawing stems from a later period; it was published after the turn of the century.)

that traditionally followed. He invoked the danger of a cholera epidemic, apparently feeling that the risk of contagion was smaller in the church than in the streets. The decision was construed by many as a patronizing interference in a popular carnival.[22] Disappointed and in-

22. The incident was fairly representative of a larger trend. The emergent bourgeoisie tried to impose its norms and values on the popular classes all over Europe. Hygienic and similar arguments often veiled moral disapproval of drinking, gambling, and promiscuity. When these pretexts were used to cancel holidays and carnivals, riots often broke out.

dignant people flocked to public buildings and started a riot. The carabinieri opened fire: They killed one, wounded several, and arrested dozens of protesters.

The next year, nearly thirty suspects were brought to trial in neighboring Bari. Their lawyer was the local positivist Giuseppe Pugliese (Toritto 1852 – Trani 1931), who also invoked collective excitement as an excuse. On the one hand, he referred to fiction: to the earlier descriptions of riotous mobs given by Hugo and Manzoni, but also to the more recent writings of Zola. On the other hand, he referred to science:

> The mutual crying, company, contact, the courage that stems from feeling united, heats the soul little by little, makes them morally drunk, makes them lose their heads without wanting to, or realizing it; they advance because they are made to, they do what the others do, the individual conscience evaporates, the feeling of individual responsibility is lost, and it is the soul of the crowd which thinks and commands, it is the body of the crowd that obeys and executes. . . . The crowds, the popular masses, form one entity, almost an organism with forces and energies, wills and sensations, ideas and actions, which is governed by laws different from those that govern individual actions.[23]

Once again, the court proved sympathetic. Pugliese expanded his plea to an article for his own *Rivista di Giurisprudenza*, and the brochure *Del Delitto Collettivo* (on collective crime, 1887). He, too, acquired local popularity and a national reputation through these and subsequent cases. And in the very next elections he, too, became a member of Parliament, where he was to remain for all but two of the next nineteen years. Yet again, science and politics proved closely related.

Although the ideas of Ferri and Pugliese were slowly winning wider acceptance, some positivists showed considerable skepticism as to the support they found in empirical disciplines such as psychology.[24] A more scientifically acceptable theory as to the why and how of these collective phenomena was clearly needed. Coincidentally, both the criminologist Ferri and the psychologist Sergi were teaching at the

23. Pugliese, *Del Delitto Collettivo*, pp. 26–8 (my translation).
24. See, e.g., Lombroso's brief review of Pugliese's *Delitto Collettivo* in the *Archivio di Psichiatria*, 8 (1887): 226.

University of Rome during these years. We have seen above that Sergi had already formulated theories on the "stratification of character" and the expression of emotions. In 1888 he opened a course on criminal anthropology with a lecture on *Psicosi Epidemica*, which later came out as a brochure, and as an article in the positivist *Rivista di Filosofia Scientifica* in 1889.

"Everybody knows," it said,

> that there are mental illnesses, and also that there are special institutions for the mentally ill. But maybe very few believe in the existence or the possibility of collective disturbances – a psychological disease propagating and spreading itself like harmful epidemics. Observations on various historical periods and on contemporary events among peoples, have clearly demonstrated to me [the existence of] such collective psychological diseases, which behave in their propagation just like any other epidemic; therefore I have named them *epidemic psychoses*.[25]

As in the case of physical epidemics, psychological contagion often spreads from a single source: usually a somewhat disturbed leader. Just as in the case of physical epidemics the external process tends to trigger an autonomous process within the various individuals themselves. Certain primitive emotions such as hatred and rage furthermore seem to spread more easily than others. The deeper mental layers, which everyone has in common, are brought into resonance, so to speak. The higher mental layers, which are only well developed in some, lose control. In this respect, Sergi also referred to the contemporary debates on mental dissolution and hypnotic suggestion, to which we will return in a later chapter.

By this time a more elaborate theory on crowd psychology was slowly emerging. Because of certain natural mechanisms, it said, the individual is thoroughly affected by the collective, and undergoes mental changes that facilitate excitement and inhibit judgment. This should be taken into account when ordinary citizens are caught in mob events, commit crimes, get arrested, and are brought to court. This thesis came to be accepted on an ever-widening scale. In early 1889, after a demonstration of a thousand unemployed workers got out of hand in Rome, the court expressed surprise that people with such "excellent antecedents" had become "violent madmen" under the influence of the

25. Sergi, *Psicosi Epidemica*, p. 1.

crowd. And after agricultural unrest north of Milan in late 1889 and early 1890, the positivist lawyer Bianchi likewise expressed surprise that the wave of riots had spread like a "psychological epidemic."

Not only uneducated workmen were subject to these processes, it appeared. In 1890 a student was tried in Naples for "resistance to the authorities." Ferri's plea as a defense lawyer was based on the arguments above and on the claim that a harsh sentence would further divide the community, whereas a mild sentence would help reestablish harmony. Another case that drew widespread attention took place in 1891 in Bologna, where radical students had been accused of "insults and rebellion." They had booed and whistled at the famous poet, professor and senator Giosuè Carducci, and disturbed his lectures. He had inaugurated the banner of the Royal Students Club, together with the controversial Francesco Crispi (who had temporarily resigned as a prime minister). Both were former rebels themselves, who had traded the republic for the monarchy and reform for the status quo.[26] In his plea, Ferri invoked the accumulated literature on criminology and psychology of the crowd. On the one hand, he pointed to criminological observations: "it is often not the guilty but the innocent who get most easily caught." And on the other hand, he used psychological observations such as "it is often because of disappointed love that people express excessive hatred." The court proved lenient.

The data on the new case were triumphantly included in the first issue of Ferri's new journal *La Scuola Positiva*. Together with the data on the cases of Bianchi and Pugliese, they were also included in the older Lombrosan journal, the *Archivio di Psichiatria*. In the latter, they formed a kind of appendix to two long articles on the subject. Somewhat later, the same material was also brought out as a booklet by the positivist publishing house of the Bocca brothers in Turin. This first monograph on the crowd was written by a young student of Ferri's: Sighele.

We have seen that Ferri and the other positivists had become more and more involved in the defense of left-wing activists and radical politics. Many of these cases were related to a major transition: from exclusively proletarian, largely anarchist inspired, relatively spontaneous, violent protest (in which "the criminal crowd" played a central role), to a middle-class-led, largely socialist-oriented, patiently

26. As a writer and a politician, respectively, they were also early forerunners of the new nationalism. See the last section of this chapter.

Part of a print commemorating the founding of the Italian Socialist Workers party in 1892. (From N. Valeri, *Giolitti*. Turin: UTET, 1972, after p. 32. Original in De Feo, *L'Italia de Giolitti*.)

organized, legal opposition (in which "party building" was the key element). In this transition, the gradual introduction of Marxist ideas played a certain role, too.

In 1891 the positivist lawyer Turati and other intellectuals of the Lega Socialista founded the theoretical journal *La Critica Sociale*. They collaborated closely with the Labor party, which also had its main base in industrial cities like Milan and to a lesser extent Turin. In 1892 the various groups founded the Party of Italian Workers in Genoa. Later, Turati traveled to the southern basin of the Po River to persuade other positivist lawyers (several of whom were friends since they studied together in Bologna) to join the party *with* the peasant federations they represented: Ferri in Mantova, Bissolati in Cremona, Prampolini in Reggio. In 1893 the organization was rebaptized as the Socialist party of Italian Workers and more than ten thousand agricultural workers and peasants expressed their adherence in an impressive march during the second congress of the party in Reggio Emilia.[27]

Whereas lawyers were overrepresented throughout Italian public life during the Postrisorgimento period, they became even more prominent in the Socialist party. Ferri even persuaded Lombroso and his son-in-law Ferrero (who was to become a noted historian) to join. A few years later, about 85 percent of the Socialist group in parliament consisted of intellectuals (in contrast with Germany, for example, where nonintellectuals played a much larger role).[28]

A number of prominent artists joined the party, too: the composer Leoncavallo, writers such as De Amicis, Graf, Negri, Pascoli, and many others. A mid-1890s poll among nearly two hundred of the best-known scholars, writers, and artists of the peninsula claimed that two thirds supported the party without reservations.[29] The historian Croce confirmed that, during this period, "Socialists, or those inclining toward and sympathizing with socialism, were no longer the few isolated individuals of the type which we described, but numbered in their ranks university students of every kind, including the most intelligent, many teachers of economics, and also teachers of law, history, and science, young men of letters, and others of an older generation to whom socialism brought a new youth."[30]

Most of the intellectuals combined revolutionary rhetoric with evolutionist scientism. Ferri, for instance, who wrote a book titled "Darwin, Spencer, and Marx," claimed that their thoughts were closely related.[31] He grew to be one of the main spokesmen for the party, traveled all over Italy, Europe, and beyond, spending more than half of his nights in a Pullman sleeper for the next twenty years according to an admiring biography by his contemporary Franchi. In the course of his close to one thousand lectures, he also became an eloquent speaker (and even involved himself in research on the psychophysiology of orators). According to his friend and colleague Ferrero, he "knew and liked the art of acting on crowds, expressing clearly what the crowds

27. Abendroth, *Sozialgeschichte der Europäischen Arbeiterbewegung*, ch. 3; Droz, *Histoire Générale du Socialisme*, ch. 5.

28. Vaussard, *Historie de l'Italie Moderne*, p. 67. Germany also had its Kathedersozialisten, of course, and intellectuals played a major role in most socialist parties, not to mention schoolteachers and their like.

29. Quoted by the German-Italian sociologist Michels in *Le Prolétariat et la Bourgeoisie dans le Mouvement Socialiste Italien*, p. 96. See also the penultimate section of this chapter.

30. Croce, *A History of Italy* (Engl. transl.), p. 148.

31. *Socialismo e Scienza Positiva* (1894), Prefazione.

already felt in a confused way."[32] In this sense too, therefore, theory and practice of crowd psychology were closely related.

Sighele and *La Folla Delinquente*

It was one of Ferri's students, then, who published the first monograph on the crowd. Scipio Sighele (Brescia 1868 – Florence 1913) was a descendant of a "grand" family of the Trentino, the main province that the Italian irredentists wanted liberated from continued Austrian rule. The family's estate was located in Nago, on the northern coast of Lake Garda just west of Rovereto (somewhat south of the town of Trento). Throughout his life, Scipio Sighele used to spend part of the summer in the mansion, which contained an impressive library. Apart from Manzoni's highly regarded *I Promessi Sposi,* which included elaborate reflections on collective behavior, it also encompassed books by a host of French authors who had published views on crowds: Hugo, the Goncourt brothers, De Maupassant, and Zola.[33]

Scipio Sighele's grandfather and father were both lawyers like himself. His father was a prosecutor in various northern provinces bordering on the "unredeemed territories," before he was appointed in Milan, where Scipio spent his high school years, and then in Rome, where he went to university. These were the years that Ferri taught penal law and "criminal sociology" there, while Sergi lectured on "criminal anthropology" and physiological psychology. Sighele's biographer Marzani says that Scipio became the "preferred student" of Ferri, and Ferri's biographer Franchi shows Sighele standing right next to the maestro on a group photograph, taken during one of their frequent excursions to prisons and asylums.

There can be little doubt that Ferri exerted a powerful influence on Sighele, not only in a scientific but also in a political sense. During these early years, Sighele showed himself keenly interested in social questions. Although he refused a "socialism of rancor," his relatives said, he did embrace a "socialism of love."[34] He remained in close touch with Ferri and other socialists throughout his life, but considered himself rather more of a progressive liberal. As we will see later, however, his real involvement was with the irredentist movement.

32. In a newspaper article quoted by Franchi, *Enrico Ferri,* p. 141 (my translation).
33. Pedrotti, Una Famiglia di Patrioti Trentini, p. 34.
34. Castellini (in his introduction to Sighele's posthumously published volume *Letteratura e Sociologia*), p. iv; Pedrotti, Una Famiglia di Patrioti Trentini, p. 6.

When still at the university, Sighele started to publish his first articles and brochures, apparently spin-offs from essays and theses he had written for Ferri. At first, he was interested in the position of women and children; this interest resulted in studies on murder and suicide in love dramas, and on abortion and infanticide by unmarried girls. Ferri introduced him to Lombroso, and from 1889 on, he started covering major trials, and doing reviews for the *Archivio* on books newly published in Italy (by Gabelli, Laschi, Lombroso, Mosso, Nocito, Pierantoni, and Vaccaro among others) as well as in France (Aubry, Bataille, Desmazes, Joly, Laurent, Moreau, Proal, Tarde, Worms, and others). The subsequent year he published an article and submitted a small thesis on the subject that was to remain at the center of his life's work in the field: *La Complicità* – in which he explored various forms of criminal complicity.

Zanardelli's new penal law (which had come into effect that same year) generally considered the association of several people in a crime an aggravating circumstance.[35] Sighele argued that this approach was one-sided. There was an obvious difference between gangs in which people involved themselves with the premeditated objective of committing crimes, he said, and crowds in which people became caught without knowing where it would lead to. The examinations committee proved skeptical, and it took Sighele several years to elaborate his theory in a more convincing form.

Meanwhile, he started digging into various categories of criminal complicity. On the one hand, he wrote articles on southern Italian gangs, emphasizing the role of biological factors. On the other hand, some of the book reviews mentioned above also focused on social influence. Among them was a book by Aubry on the role of "contagion" in murder, for instance, and a book by Tarde that emphasized the role of "imitation" and "suggestion" in crimes and also mentioned those crimes committed by crowds. Gradually, the latter problem became Sighele's primary concern. Somewhere in the first half of 1891, he published his two articles on the subject. These articles formed a monograph that was immediately reprinted as a small book. Sighele's title *La Folla Delinquente* (the criminal crowd) implicitly referred back to Lombroso's title *L'Uomo Delinquente*. But it was often misread (particularly in France and other foreign countries) as implying that crowds were always criminal. Therefore the title was later changed

35. In its article 248, among others.

again, which – along with the language problem – contributed in no small measure to the confusion over various editions of this book and subsequent ones in the historical literature.[36]

The original edition of Sighele's book comprised an introduction and three chapters. In the introduction he discussed the relationship between psychology and sociology, and he leaned heavily on Spencer, who was widely considered to be *the* founding father of the new science of individual and society at the time, and whose works had recently been translated into both French and Italian.[37] It has been established, he said, that the physical nature of the elements also determined the possible nature of their chemical reactions. In the same way, it added, the psychological nature of people determines the possible nature of their social relations. This should not be taken to mean, though, that the psychology of the individual was identical to that of the collective. According to Sighele, experiences with court juries, scientific commissions, and political meetings indicated that people reacted differently when isolated or united with others. This was especially true for casual encounters of large groups of people from different backgrounds in crowds. Next to individual psychology, then, there was a definite need for the new discipline of collective psychology – which was to study the behavior of people in such large groups.

The first chapter of the book opened with a few remarks on penal responsibility but was devoted to "the psychophysiology of crowds." Not only animals, Sighele said, but humans too, show a clear tendency to adopt behavior already displayed by others. By now there existed quite a range of literature on theft and murder committed by several people in similar ways. Three explanations had been offered: moral contagion, social imitation, and hypnotic suggestion.[38] Two others should be added, according to Sighele. First an "anthropological" factor: primitive (emotional) tendencies were more easily stirred by the crowd than civilized (rational) ones. And second a "numerical" factor: The larger the number of people expressing a certain emotion, the

36. Ferri's April plea contains no reference to it; Garofalo's review was officially published in August.
37. Sighele quoted from the French editions of the *First Principles*, from his *Psychology* and his *Sociology*. Note that the latter had been translated by Espinas and Ribot. A fuller discussion of these ideas is in the chapter on Le Bon.
38. Part of the background of the idea of moral contagion was discussed earlier in this chapter. The backgrounds of the ideas of hypnotic suggestion and of social imitation will be discussed in the last two chapters of this study.

more intense the feeling would become. The result of all these processes added together was the development of a kind of mental unity in the crowd, he said, the development of a kind of collective soul.

The second chapter thereupon provided a large number of concrete examples from fictional and historical literature.[39] It opened by stating that it was understandable that the persistence of miserable conditions among the popular classes, together with the rise of new political ideas, had triggered mass protests that sometimes got out of hand. People got excited, especially when their long-felt powerlessness turned into a sudden experience of temporary power. Not only collectives, but individuals too, had been known to commit excesses under such circumstances. At the same time, it should be acknowledged that some people were better at resisting these temptations than others.

After these psychological observations, the third chapter turned to the judicial conclusions to be drawn from all this. Earlier in the book Sighele had pointed out that through the ages the question of the attribution of penal responsibility to people involved in criminal crowds had received unsatisfactory answers. Roughly two approaches could be distinguished. One approach held all individuals caught in such a situation equally responsible for the events. This could mean society had to punish everyone equally. A variation was to punish a fixed proportion severely and to let others off the hook (as in "decimation" – a method often applied after mutinies). The drawback of both was that relatively innocent people were made to suffer along with the really guilty. When applied to riots, furthermore, this approach usually meant that people caught by mere coincidence were held responsible. The drawback here was that naive people were caught more easily than cunning ones, so that one ended up, in fact, punishing the wrong individuals. Therefore an alternative approach tried to distinguish between people held more responsible and those held less responsible for the events.

One could, for instance, try to catch the "instigators": those who had stirred up enmity or pointed out targets. Yet these people had often not been involved in the excesses themselves and had just expressed opinions or proposed actions. One could also try and catch the "executers" of the crimes: those who had participated in the destruction of

39. Oft-quoted authors in the second edition of the book were on the one hand Italian positivists such as Ferri, Lombroso, Pugliese, Sergi, and Garofalo; and on the other hand French theorists such as Tarde, Taine, Espinas, and Ribot. Novelists such as Manzoni and Zola were also quoted frequently. Spencer was an oft-quoted non-Latin author.

property or the maltreatment of persons. Yet this actual participation was often hard to prove, and even so it was often hard to say to what extent their gestures had contributed to the actual looting or lynching. Was it the first and tentative probe that counted, or the last and decisive blow? Was there a difference?

Neither solution seemed practical. The reason, Sighele and the other positivists felt, was the confusion between responsibility and imputability (a distinction subsequently adopted in some legal idiom but hardly copied in general language). It was related to the unscientific idea of moral guilt, they felt, and of a fixed penalty for certain offenses. Instead, one should look at the problem from the perspective of social defense. One should try to evaluate the threat that the criminal posed to society and the risk that he would commit the same acts again. Unfortunately, Sighele concluded, the present law hardly allowed for such distinctions.

Sighele's book received rave reviews from his colleagues. Not only in *La Scuola Positiva,* which successively reviewed it on the occasion of its first issue (a "triumphal start . . . of a scientific career"), its first foreign translation (a "novel and true" theory), and its second edition (a "work of genius").[40] But it was also favorably reviewed by Pugliese in the *Rivista di Giurisprudenza* ("an excellent book"), and by Garofalo in the *Tribuna Giudiziaria.*[41] The latter added, though, that Sighele had not been entirely consistent in his "positivist" approach. He suggested that judges apply the law in all its rigor to born and habitual criminals who had committed crimes under the cover of the crowd – since they had only followed their inclinations. The same judges, however, might take a more lenient attitude when occasional or passional criminals were caught in such a situation, since in that case their behavior was the obvious result of mental confusion or irresistible force or both. Sighele decided to embrace these conclusions, and included them in subsequent editions of his book.

Other reviewers in *La Giustizia* and the *Archivio Giuridico* were laudatory, too, but felt Sighele could have paid more attention to the social groups from which these turbulent crowds were usually recruited, and should have more clearly acknowledged that even people with apparently bad backgrounds might often display surprisingly

40. 1891, p. 177; 1893, p. 427; 1894, p. 995, respectively (my translation).
41. Both in 1891.

good behavior.[42] Similar criticisms were expressed after the French translation of 1892 (see the next chapter), and by the Italian socialists. Their theoretical journal *La Critica Sociale* had originally called the book "one of the best works published in recent years by the positivist school," but had returned to the question a few years later by carrying an extensive discussion between Tarde, Sighele, and Ferri on the merits of collective behavior.[43] Sighele had claimed that the crowd was always more emotional and less rational than the individual, but Ferri countered that quite often collectives proved not only braver but also wiser than isolated people. Of course this was a basic tenet of socialist thought, and it is not surprising that other leftist crowd psychologists should later take up this same issue again.

Meanwhile, Sighele's theory continued to be widely invoked to exonerate ordinary people caught in social unrest. A strike in a hat factory, to give one example, had led to disturbances and arrests. But at the trial in Pallanza, the defense invoked crowd psychology, so that nineteen suspects were freed and five came off with relatively light sentences. A strike of alabaster workers, to give another example, had led to a similar confrontation in Volterra. Here again, Ferri and the other lawyers succeeded in getting lenient treatment for the seventeen suspects, by referring to the "new scientific discoveries."

The tide was slowly changing, however. Now that trade unions and political parties became organized on a wider scale, strikes and demonstrations rapidly grew in size, so that related disturbances posed an increasing threat to the Postrisorgimento order. This became obvious as the nineties progressed, and two major waves of social unrest created quasirevolutionary situations.

The first of these centered on Sicily, which had remained the poorest and most exploited region of the realm. Workers and peasants had formed primitive organizations called *fasci* (a term meaning "groups" and often depicted by a bundle of sticks: a symbol of social authority stemming from Roman antiquity and revived during the French Revolution). Their social offensive had led to widespread unrest and frequent confrontations with landowners and local authorities. In early 1894 the government decreed a state of siege and sent in the army to quell the unrest. Thousands were arrested. The procurator general in Palermo reported: "During the past few days we have seen ferocious

42. Vol. 2, No. 35; and Vol. 47, No. 5, respectively. On other translations, see note 48.
43. 1891, 1894, respectively. The discussion was included in *La Folla Delinquente* from the second edition on.

mobs move against private houses and public buildings; crowds of unarmed women, workers overtaken by mad delirium, and guided by wicked agitators, ravaged, sacked, started fires, and gave themselves over to deplorable excesses. Entire regions fell prey to revolt and anarchy; riots and rebellions were so impetuous that – after all other forms of pacification had been tried in vain – violent repression became inevitable."[44]

The procurator, who made thinly veiled references to the "laws of collective psychology," was the same person to whom *La Folla Delinquente* had been dedicated: Scipio Sighele's father. The Sicilian events also contributed to a temporary ban on the Socialist party. The climate quickly deteriorated.

The second edition of *La Folla Delinquente* was published in 1895 and also received much praise. By that time, Sighele Jr. had decided to capitalize on his newly acquired fame, and make the book into one of the centerpieces of a multivolume work on all aspects of crimes committed by more than one person. He elaborated his earlier minor thesis, *La Complicità;* it became a more solid study, and the "judicial part" of the series. He also published another widely acclaimed study on the simplest form of criminal cooperation, *La Coppia Criminale* (the criminal couple) and dedicated it to his mentor Ferri.[45] Furthermore, he projected a volume on *Le Associazione di Malfattori* (associations of malefactors). It was never completed, though the theme of criminal gangs and organizations ran through several overviews which he published with close colleagues in these years.[46] And finally, there was a study on the criminality of sects: *La Delinquenza Settaria* (sectarian criminality), published in 1897. This book borrowed from an article by his French colleague Tarde, to which we will return.

It extended the previous line of argument from crowds to sects – secret clubs locked in violent combat with the established order. He tried to demonstrate that the two forms of collective behavior were

44. The report was reprinted as Appendix 2 to Pedrotti's long article on the Sighele family (my translation).

45. Ferri invoked the study in a famous court case in 1902, when he defended Tullio Murri, who had killed the husband of his sister Linda. The Murris were children of a well-known socialist physician. Sighele assisted the family of the victim, Conte Francesco Bonmartini, who was a noted clerical conservative. See Garbari, *L'Età Giolittiana nelle Lettere di Scipio Sighele*, pp. 84–5, 113.

46. *Mondo Criminale Italiano*, Vols. 1, 2 (with Bianchi and Ferrero, 1893, 1895); *Cronache Criminali Italiane* (Vol. 3, with Ferrero, 1896); *La Mala Vita a Roma* (with Niceforo, 1899).

closely related: one being temporary and heterogeneous, the other permanent and homogeneous (in the sense that it had a selected membership). The problem of criminal responsibility was almost the same, Sighele said.

> The man who, inebriated by the cries of the multitude surrounding him, strikes or kills, is from the psychological point of view rather similar to the sectarian who – under the suggestion of ideas which he had heard develop in secret societies – takes a shot with a revolver at the real or supposed tyrant, or stabs him with a knife. Both are only the *executors* of a crime which has been *thought out* by others; they are simple *automatons* carrying out the wishes of another *mind*. Between them, there is only a difference in degree. The former is the victim of an *immediate, static,* and therefore *unconscious* suggestion; the latter is the victim of a *mediate, dynamic,* and therefore *less unconscious* suggestion. The responsibility of the second will therefore always be greater than that of the first; but one should not forget that the true author of the crime committed by the first is a collective entity – the crowd; and that in the same way the true author of the crime committed by the second is also a collective entity – the sect.[47]

Sects are the *trait d'union* between crowds and associations, Sighele said; the propagation of most major religions and ideologies had started with such minorities. So, like crowds, sects were an inevitable part of social life. Apart from this extension of collective psychology, however, there was also a political twist to the book, to which we shall return.

During these same years, Sighele also wrote an increasing number of articles with a more general literary or philosophical strain for journals such as *La Nuova Antologia*. Several of his major works were by now translated into French, and *La Folla Delinquente* was eventually translated into no less than a half dozen foreign languages.[48] Sighele had become a noted criminologist, respected both at home and abroad. Whereas he had been unable to present his pioneering ideas on the crowd to the Second International Congress of Criminology (for reasons we will discuss in the next chapter), recognition finally came to him at the Fifth International Congress in Amsterdam

47. Chapter III-4, p. 139 in the French ed. (my translation; emphasis in the original).
48. French, German, Spanish, Russian, Polish, and Dutch.

in 1901. He was not only invited to present a paper titled *Le Crime Collectif,* but his ideas also permeated contributions by other specialists.

During these years his mentor Ferri made several speaking tours through the Netherlands and particularly Belgium, advertising collective psychology as well. Every year, for instance, he lectured a few weeks as a visiting professor at the unofficial and progressive Université Nouvelle in Brussels, which in the midnineties had broken away from the official and more conservative Université Libre.[49] He also introduced Lombroso and other representatives of the Italian school of positivist criminology, including Sighele, to the university. There, the latter taught a course in collective psychology every year of the first decade of the twentieth century. Some of his notes for these lectures, and for others given at "popular universities" elsewhere, survive today.[50]

But Sighele's attempts to create a similar position at one of the "free schools" in Paris were less successful. In Italy he often lectured for "popular universities" and similar institutions, but only briefly held minor positions in the academic world.[51] However, the Fifth International Congress of Psychology, chaired by Sergi in Rome in 1905, did include a section on criminal psychology to which Lombroso, Ferri, Sighele, and many others contributed. There were ample discussions on collective psychology, as well. One of their participants was another Italian scholar who had made a major effort to get the new discipline accepted: Rossi.

Rossi's crowd psychology and the crisis of 1898

During the late nineties, another wave of riots gave a major impulse to the further elaboration of collective psychology in Italy. Grain prices had risen sharply: first as a result of bad harvests throughout Europe, later because of the Spanish-American War. In late 1897 and early 1898, riots became increasingly frequent at bakeries and merchants' warehouses: first in Sicily, then Puglia and Calabria, later in the

49. The new university later amalgamated with the Institut des Hautes Études de Belgique. See Goffin (1969), Despy-Meyer (1973), and Despy-Meyer and Goffin (1976).

50. Mrs. Garbari located these notes in the Risorgimento museum of Trento.

51. According to some sources, he lectured at the universities of Rome and Pisa from 1899 to 1902.

Encounter in 1898 between demonstrators and carabinieri on the Corso Venezia in Milan, as seen by *L'Illustrazione Italiana.* (From G. Talamo, *Storia d'Italia,* Vol. 4, Pl. IV. Original in *L'Illustrazione Italiana,* May 15, 1898.)

Marches, Tuscany, and Emilia-Romagna. In May 1898 the unrest reached the industrial region of Lombardy. After barricades sprang up in the city of Milan, General Beccaris employed cannons to restore order. During these months, hundreds of peasants and workers were killed. In the end, thousands of activists were arrested, including Turati and several other leaders of the Socialist party.

Ferri was one of the few leaders spared, and he suddenly became the main spokesman for the party.[52] Local committees for the liberation of political prisoners were formed. One of these committees was led by Ferri's old acquaintance Rossi, who felt that collective psychology was to blame for some of the atmosphere. He felt Sighele's undue emphasis on the criminal inclinations of crowds had provided the authorities with a perfect excuse to quell mass protests. Someone was bound to elaborate a collective psychology more in line with Ferri's earlier remarks in *La Critica Sociale,* which stressed that crowds did have a social value, too.[53]

Pasquale Rossi (Cosenza 1867 – Tessano 1905), the son of a lawyer, had briefly studied law in Rome where he may have met Ferri.[54] He left soon thereafter, though, to take up medicine in Naples (where he was once arrested during a workers' riot). After completing his studies he returned to his native town, and started an outpatient clinic, where poor patients were treated free of charge. He also set up a local chapter of the Socialist party, and was its representative at the Reggio-Emilia and subsequent congresses. Somewhat later, he got into trouble with the regional authorities for publishing a series of letters of complaints to party leader Turati. Apparently this brochure, *I Perseguitati* (the persecuted), was his first attempt at analyzing a social phenomenon from a psychological point of view. He was elected a member of the municipal council during these years, but subsequent campaigns for a provincial and even a national mandate failed.

The events of 1898 had convinced him that it was of the utmost importance to elaborate a collective psychology with a more balanced approach. During the next seven years, he published no less than seven books in the new field: on the soul of the crowd (1898); collective psychology – studies and research (1899); mystics and sectarians – studies in collective psychopathology (1899); Giuseppe Mazzini and modern science – studies in sociology and psychology (1899); morbid collective psychology (1901); the "suggesters" and the crowd – psy-

52. He became editor of the party newspaper *Avanti* and leader of the parliamentary group. The new prime minister, General Pelloux, first tried to calm things down but in 1899 resorted to a hard line by decreeing emergency laws limiting the freedoms of the press, assembly, and association. Ferri led a filibuster campaign. New elections brought large gains for the extreme Left (which scored a total of a hundred seats). Soon thereafter, an anarchist emigrant returned from the United States, killed the king and "took revenge" for the Milan massacre.
53. This can be inferred from an unpublished outline for a new preface to his book *L'Animo della Folla,* quoted by his friend and colleague Squillace (1909, p. 27).
54. According to Squillace (ibid.), this was during the academic year 1885–6. Ferri began to lecture in Rome in 1886.

chology of leaders (1902); and finally on sociology and collective psychology (1904). In these books, he continued in the true tradition of Lombroso, Ferri, and Sighele, but attempted to go even further. He tried to incorporate some ideas of Le Bon and Tarde, but also to steer clear of one-sided emphasis on the negative aspects of crowd behavior.

He proposed to make a clear distinction between social psychology, which he identified more or less with the psychology of peoples, and collective psychology proper, which involved people from varying backgrounds. All the same, he said, collective phenomena should also be seen in a proper sociohistoric perspective. Of course they differed in advanced and backward regions, such as northern and southern Italy, just as they differed in various epochs, such as modern times, the middle ages, or antiquity.

One should understand that such mass phenomena were stimulated by social contrasts and came in two forms. On the one hand there were short-lived, unstable, undifferentiated (physically assembled) groups such as crowds. They were often highly emotional, but not exclusively delinquent; theatrical, religious, and military meetings showed that they might well be inspired by noble feelings. On the other hand, there were longer lasting, stable, differentiated (physically dispersed) groups such as sects, movements, and classes. They often contributed to cultural life: to folklore, language, literature, art, and other expressions of culture.

All collective phenomena had a natural history and were cyclical in nature: They had "juvenile" and "senile" phases. Crowds had a certain typical rhythm, too (which he felt was related to respiration). Within crowds, one could usually distinguish an amorphous majority and an active minority, furthermore, the indirect and direct leaders. Most important of all, crowd psychology should move from theory to practice, by involving itself in the moral and intellectual education of the masses, and the spreading of mass culture (*follacultura*).

Rossi himself tried to promote this program through a new journal, entirely devoted to the new discipline: the *Archivio di Psicologia Collettiva e Scienze Affini* (archives of collective psychology and related sciences). It was published in his hometown Cosenza, in spite of the fact that "the long silence of the masses" in Calabria seemingly made that region ill-suited as a "laboratory of collective psychology."[55] The first number was issued on April 1, 1900 – the first Fool's Day of the

55. Anno I, Fasc. 1, p. iii.

new century and therefore a somewhat ominous date. It listed well-known (prospective) collaborators such as Ferri and Sighele; Le Bon, Tarde, and later "Durgkeim" (!) and others.[56] But articles came mainly from minor authors with an active interest in the subject, such as Groppali, Squillace, and of course Rossi himself. This might be one reason why the journal disappeared again, somewhere in its second year of publication.[57]

Another socialist author was the early psychologist Paolo Orano (1875–1945). His book, *Psicologia Sociale* (1902), was partly a collection of articles that had already been published in journals. It was the first book in Italy with this title, but it claimed the new interdiscipline had already been foreshadowed by Carlo Cattaneo and his *Psicologia delle Menti Associate* (1859). Orano's book was a violent polemic against the surviving spiritualist and deductive approach to the field, and a plea for a materialist and inductive one. Rather than proceeding from ideas to things, he said, one had to proceed from things to ideas (p. 207). Furthermore, individual psychology was severely limited because "Man becomes man only in, and for, society. . . . There is no human psychology outside the complex of human relations . . . Just as the cell has its particular function insofar as it participates in the substance of an organism, so too does the individual have his particular functions, characteristically dispositional, insofar as he participates in society."[58]

Orano felt the psychologist should study sociological and biological facts: They revealed Marx's laws of history and class struggle as well as Darwin's laws of evolution and the struggle for existence. Social and collective psychology covered two important fields: the psychology of crowds and that of peoples. The book referred to crowd scenes from famous novels by Manzoni but also Balzac, Hugo, and Zola; and to scientific studies by Lombroso, Ferri, Sighele, but also Le Bon and Tarde (to whom an entire chapter was devoted, alternately laudatory and critical). Many years later Orano wrote a preface to another review of the criminological literature in Manci's book *La Folla* (1924).

56. Orthographic and typographic errors abound in Rossi's journal and early books: it was apparently hard to find a suitable printer and/or publisher in Cosenza.

57. I have not been able to find more than five issues of the first year, but Rossi's *Sociologia e Psicologia Collettiva* (p. 90) contains a reference to the sixth issue of the second year.

58. Pp. 76–82. English translation from Gregor, *Young Mussolini*, p. 60. Gregor's third chapter elaborates on the link between Orano, the syndicalists, and fascist ideology.

The other main field was the psychology of peoples or *demo-psicologia,* according to Orano. There was a great need for a better understanding of what constituted the national soul: *l'italianità* (p. 250). This was essential to achieve further unification and a revival of the country. In fact the last sentence of the book said that, while writing, he felt speaking within him, the "most healthy and resounding voice" of the two and a half times millenary political spirit of Italy.

Orano joined the revolutionary wing of the Socialist party (which was now headed by Ferri) in criticizing party leader Turati (who took an intermediary position) and the reformist wing (now headed by Bissolati), for being overly eager to control spontaneous outbursts of mass agitation. Together with other syndicalists, Orano began to emphasize that all human progress was the result of the violent pursuit of collective ideals. His syndicalist comrade Sergio Pannunzio invoked the "scientific laws" to collective psychology in advocating insurrectionist policies.[59] On the occasion of the first general strike, the like-minded Arturo Labriola summarized this same point by saying that "five minutes of direct action were worth as much as many years of parliamentary chatter."[60] Similar opinions were voiced simultaneously in France by Georges Sorel (who incorporated several ideas from Gustave Le Bon) and in Germany by Robert Michels (who incorporated several ideas from Max Weber and the Italian elite sociologists).[61] It was, in fact, such debates that inspired Sorel's book on collective violence and Michels's book on party oligarchization.[62] But whereas the radical Italian intellectuals were closely involved in the mass movements of their day, their French and German colleagues often remained fairly isolated. Thus, both Sorel and Michels often sought the company of their Italian comrades during these years (and even published some of their studies in Italian first).

The same atmosphere of impatience also pervaded part of the Italian socialist press, including one particular bi-weekly that had adopted the proud name *La Folla.* The first issue explained: "Ours is a virile mass [crowd], which moves and acts, which unites itself and cries out,

59. Orano later became one of the founders of the Fascist party and press (and even inspired some of its racism). For more on Orano, see ch. 5, by Doise, in Graumann and Moscovici, *Changing Conceptions of Crowd Mind and Behavior.* For Pannunzio, see Gregor, *The Ideology of Fascism,* p. 112.

60. Salomone, *Italy from the Risorgimento to Fascism,* p. 51.

61. See Kautsky and Luxemburg, f.i., in Grunenberg, ed., *Die Massenstreikdebatte.*

62. See Mitzman, *Sociology and Estrangement.*

whenever it is denied a right by the law of privilege."[63] The Milan magazine appeared from 1901 until the epoch-making general strike of 1904, and again from 1912 until Italy's entry into the war in 1915. It was published by the leftist journalist Paolo Valera, who also wrote a novel with that same title *La Folla.* Among its collaborators were Enrico Ferri as well as a close friend of many revolutionary syndicalists, who signed with the French pen-name "l'homme qui cherche," and who was none other than Benito Mussolini.[64]

Sighele and the Associazione Nazionalista Italiana

In order to present things in their proper perspective here, we need to make a brief excursion to another field, complementary to crowd psychology: the sociology of elites. While the socialist movement continued to expand, liberal thinkers also came to challenge the premises of parliamentary democracy. At the time of the provisional completion of unification, illiteracy still stood at 70 percent, and only diminished with some 10 percent per decade thereafter. In 1882 the electorate was extended from one-half million to two million voters by changing the financial and educational requirements,[65] but politics continued to be dominated by local notables. Furthermore, successive prime ministers had blunted the rise of an effective opposition by trading regional favors for parliamentary support. This ingenuous maneuvering often bordered on outright corruption. During the early nineties, a major scandal developed, with indications that the Banca di Roma had bribed politicians to overlook a massive overemission of bank paper and similar improprieties. This affair further contributed to public skepticism about the nature of representative government.

While some lower and middle class reformers in the socialist and radical parties had developed theories on the emotionality and irrationality of masses and crowds, some middle and upper class activists in the liberal and conservative movements had meanwhile developed more or less complementary theories on the inevitability of elite rule. Although they came from somewhat different political horizons, the early crowd psychologists and elite sociologists were vaguely familiar

63. Anno 1, No. 1 (May 5, 1901), (my translation).
64. Notizia Biografica in Valera, *La Folla*, p. 27.
65. Complete statistics on the gradual extension of voting rights in Ghisalberti, p. 219.

with each other's work,[66] and also shared a secular ideology in which individual inequality and social evolution formed key elements. The latter's ideas fit into a long Italian tradition, going back to Machiavelli (political amorality), Vico (historical cyclicity), and various Risorgimento thinkers.[67] They, too, advocated the empirical study of political behavior rather than a vocal adherence to high-minded ideals. Such an unprejudiced investigation, they said, would bring to light the fact that in spite of all the mass rhetoric, elite rule had been a persistent reality in all cultures and epochs.

In spite of the priority debate in this field, too, there can be little doubt that the first to formulate these ideas in an elaborate form was Gaetano Mosca. As a law student in Palermo he had already been fascinated by the implications of Taine's *Ancien Régime*. In the course of the eighties, he published successive studies on national identity, parliamentary government, and modern constitutions, which claimed in various forms that unorganized majorities could never exert power, but that organized minorities or individuals could. "It is not the voters who elect the deputy," he said in one famous phrase, "but ordinarily it is the deputy who *has himself elected* by the voters."[68] His shrewd observations on the political process brought him to the national capital as an editor of the journal of the Chamber of Deputies, and as a protégé of the Marquis di Rudini: twice a prime minister and a long-time liberal-conservative leader (whom Mosca would even later succeed as a deputy).[69] Mosca's Roman experiences confirmed his early intuitions and led to his most mature and well-known work: *Elementi di Scienza Politica* (1895–96).

He opened by pointing out that "During centuries past, it has many times occurred to thinkers to consider the hypothesis that the social

66. Mosca, originally a lawyer, was familiar with Lombroso's theories. He lived in Rome at the time when both Sergi and Ferri lectured (and Sighele studied) there. Pareto reviewed books of Lombroso, Ferri, Sighele, and Rossi, but was more directly influenced by the French authors Le Bon and Sorel. The key books by both Mosca and Pareto contain many implicit and explicit references to crowd psychology.

67. Ripepe, *Le Origini della Teoria della Classe Politica*. On the history of elite theory in general, see Bachrach (1970), Bottomore (1964/1973), Burnham (1943/1970), Sereno (1968), and others.

68. Mosca, *Sulla Teorica* etc., p. 295 (transl. in Salomone 1945, p. 19). It is obvious that this held particularly for nineteenth-century Sicily.

69. Mosca (Palermo 1858 – Rome 1941) was a deputy from 1909 to 1919, a senator from 1919 to 1925, and a vice minister (for colonies) from 1914 to 1916. Also see Meisel (1956, 1965), Vecchini (1968), Burnham (1970), Bobbio (1972), Albertoni (1973), Nye (1977).

phenomena unwinding before their gaze might not be mere products of chance . . . but rather effects of constant psychological tendencies determining the behavior of the human masses." He continued by stating that "Among the constant facts and tendencies that are to be found in all political organisms, one is so obvious that it is apparent to the most casual eye. In all societies . . . two classes of people appear – a class that rules and a class that is ruled." These groups are being constantly renewed. "In the first place . . . within the lower classes another ruling class [group], or directing minority, necessarily forms. . . . In the second place, whenever and wherever a section of the ruling class tries to overthrow the legal government . . . it always seeks the support of the lower classes, and these readily follow its lead."[70] The claim that the socialist movement could put an end to elite rule was therefore patently false. Mosca's provocative thesis brought him national fame as a political scientist, and a professorial appointment in Turin – where he befriended Lombroso and his pupils.[71]

At the behest of their mutual acquaintance Maffeo Pantaleoni, the book was sent to another Italian professor,[72] who was just then embarking on similar ideas. This Marquis Vilfredo Pareto was an engineer who had become a company director in Florence and a major spokesman for free trade in the *Giornale degli Economisti*.[73] Once having made an unsuccessful attempt to get himself elected deputy, he and Maffeo Pantaleoni later played a key role in exposing other deputies' involvement in the Banca di Roma scandal.[74] He was happy to leave the Italian scene shortly thereafter, by accepting a chair of political economy at the Swiss University of Lausanne. His first book on that subject contained the nucleus of his sociology. But after receiving a large inheritance, he finally found the opportunity to settle down and write a more complete version.

In an Italian article published in 1901, and then in a French book published in 1902–3, he reformulated and elaborated the theory of the "circulation of elites," as he called it, writing,

> The phenomenon of elites which, through an incessant movement of circulation, rise up from the lower strata of society, mount up

70. Mosca, *The Ruling Class* (English translation), pp. 1, 50 and 117–8, respectively.
71. Particularly Ferrero, Niceforo, and Michels, who all formulated theories related to his.
72. See Meisel, *The Myth of the Ruling Class*, pp. 175–7.
73. On Pareto (Paris 1848 – Céligny 1923), see: Meisel (1965), Busino (1967), Tommissen (1967), Freund (1974), and Nye (1977).
74. Details in Thayer, *Italy and the Great War*, pp. 57–74.

to the higher strata, flourish there, and then fall into decadence, are annihilated and then disappear – this is one of the motive forces of history, and it is essential to give it its due weight if we are to understand great social movements. Very often the existence of this objective phenomenon is obscured by our passions and prejudices, and the awareness we have of it differs considerably from reality.[75]

The book, *Les Systèmes Socialistes,* set out to prove that the movements' theories were irrational and sentimental and only catered to the unrealistic and wishful thinking of the lower classes.

Meanwhile, the elite sociologists were not the only ones to question the functioning of democracy. Some of the early crowd psychologists did, too, as illustrated by the case of Scipio Sighele himself.

I mentioned that Sighele's ideas took a peculiar turn in the second half of the nineties. In certain respects, this turn paralleled the thinking of the authors mentioned above, although it is probably incorrect to simply call him an elitist himself, as one of his closest family members did.[76] It is true, however, that he launched a violent attack on parliament, and on politics in general, during this period. This may have had various immediate reasons, one of them being the transfer of his father from Sicily for having dared to link the murder of a bank director to a corrupt deputy.[77] Soon thereafter, Sighele Jr. wrote his pamphlet *Contro il Parlamentarismo,* which was published in 1895. It argued that the representatives did not behave like a national elite, but rather like an ordinary crowd. Part of the reason, he said, was that the Chamber was too large – which made it function just like any other mass meeting, and invited demagoguery.[78] Many felt, though, that he was attacking the democratic system as such. A long polemic ensued, in which people like Ferri and Bissolati, but also Mosca and others, participated.[79]

75. From the introduction to *Les Systèmes Socialistes.* Translation in Finer (1976), p. 134.
76. That is, his younger nephew and close friend Gualtiero who, however, probably projected his own feelings onto his uncle here. See Castellini in Sighele (1914), p. iv.
77. He had found out that the murder of the Marquess di Notarbartolo, senator and director of the Bank of Sicily, had probably been ordered by deputy Palizzolo, for having revealed certain politicofinancial transactions. See Garbari (1977), p. 26, n. 34.
78. The German *Kleine Brockhaus Enzyklopädie* of 1910 (Vol. II, p. 356) listed the size of various parliaments existing throughout the world at the time. According to this list, Italy's chamber numbered 508 members, Austria 516, France 584, and Great Britain 670. Most of the others were considerably smaller (Holland, for instance, had only 100).
79. A violent and prolonged polemic was fought out with Bissolati in the pages of *La Critica Sociale* (1895–96).

The subsequent year Sighele elaborated his viewpoint in another pamphlet, *La Morale Individuale e la Morale Politica.* In it he noted that there was an obvious contrast between the moral standards that people seemed to apply to individual behavior, where theft and violence were considered unacceptable, and those they applied to political behavior, where corruption and repression were considered normal. A year later, these long essays were included in his new book on the criminality of sects, which in a later edition was even retitled *Morale Privata e Morale Politica.* The book also stated that there was an obvious contrast between the "primitive" crimes of the poorer classes, such as the violent excesses of crowds and sects, and the supposedly "civilized" crimes of the rich and powerful, such as sophisticated malversations of bankers and politicians.

This sudden downgrading of parliamentary politics coincided with an abrupt upgrading of mass movements, and a more marked acceptance of the positive role of collective behavior in historical progress, incorporating some of the ideas of Ferri, Rossi, and others. This is not only clear from a number of articles that he wrote and published during these years, but also from the gradual metamorphosis which one of his oldest works underwent: *La Folla Delinquente.*

As was usual in his circle, new material was included in new editions. Thus the second French edition of 1901 included not only data on a number of new court cases but also new chapters on positive social phenomena such as art, morality, and public opinion. By then the book had expanded from its original mere ninety pages to some three hundred pages. The third Italian edition of 1902 was therefore split in two and retitled – something that further contributed to the subsequent confusion in the foreign literature.[80]

The first volume was limited to the original subject and called *I Delitti della Folla* (the crimes of the crowd, 1902/1910). Note the obvious intention was to undo the false impression created by the original title that crowds were inherently criminal. The second volume included all the new material on the positive aspects of collective behavior, and was called *L'Intelligenza della Folla* (the intelligence of the crowd, 1903/1922). In order to stress the point, it also included another chapter that was quite revealing of Sighele's change of mind. Whereas many of his positivist colleagues were discussing the supposed superiority of the northern, Germanic peoples, and the supposed

80. See note 13 to the introduction of this study.

inferiority of the southern, Latin ones, Sighele chose to hail the romantic nationalism of the writer Gabriele D'Annunzio, and its contribution to a revival of Italian pride.

Once again, this development must be seen in the wider context of the evolution of Sighele's own political preoccupations. I mentioned before that his family originated from the Trentino province, that he used to spend part of the year at the family estate in Nago (on the Austrian side of Lake Garda), and that he had increasingly involved himself in the irredentist movement, which wanted these territories liberated from foreign occupation. An Austrian-Hungarian census of this period acknowledged that 97 percent of the population of the Trentino spoke Italian as its "habitual" language, at least 62 percent in the Trieste district did, 49 percent in Fiume (Rijeka), and some 37 percent in Istria and Gorizia-Gradisca.[81] As a well-known scholar, Sighele had become a frequent speaker and a major mentor for associations of students and mountaineers in the Trentino.

One of their major ongoing battles was for the establishment of Italian faculties or universities in one of the major Italian-speaking cities of the region, such as Trento or Trieste. The Austrian government, however, feared that these would immediately become staging grounds for separatist agitation, and proposed the German-speaking capital of Vienna instead. After several years, the provincial capital of Innsbruck emerged as a possible compromise. For some time, Sighele even tried to set up an alternative-style university there, modeled on the Université Nouvelle in Brussels, and to enlist the help of many of his colleagues. Gradually, he became one of the main spokesmen for the irredentist movement and a noted member of the national associations "Trento and Trieste" and "Dante Alighieri."

At the time, nationalism was slowly becoming a common meeting ground for activist intellectuals from both the Left and the Right. It was a shame, many felt, that – for lack of economic opportunity – hundreds of thousands of Italian citizens were every year reduced to emigration, and to joining the building of other nations rather than their own, whether in Southern Europe, Northern Africa, or the Americas. They also said it was a shame that Italy was not able to provide them with adequate protection abroad, or to enlist their energetic labor for the creation of a colonial empire – as Great Britain, France, and

81. Quoted in Seton-Watson, *Italy from Liberalism to Fascism*, p. 353.

even minor countries had done. More decisive military action should therefore be taken, particularly in those parts of eastern and northern Africa that had not yet been occupied by other European powers. One should not be satisfied with the present-day mediocre *Italietta,* as they scornfully called it, but undertake to build a new Roman Empire instead.

These sentiments received an even wider echo after 1908 when the revolt of the Young Turks sounded the disintegration of the Ottoman Empire, when Austria-Hungary moved to annex Bosnia-Herzegovina, and when unrest spread through the eastern Mediterranean. Scipio Sighele decided to go on a tour through various Balkan countries, whereas his young nephew Gualtiero Castellini (who was almost like a son to him) won a trip to Libya awarded by the Naval League. In 1910 they were closely involved in the founding of the Associazione Nazionalista Italiana. It had been preceded by a three-day conference in Rome, with Sighele as chairman and Castellini as secretary, and with several hundred participants from varying backgrounds.

Sighele gave a predictable speech on the relation between nationalism and irredentism, but another major speaker came up with an entirely different theme. It had first been formulated by the ex-socialist Pascoli, but was now taken over and developed further by the nationalist Corradini. Whereas socialist internationalism divided the country, he said, a "national" socialism would unite it. Italy should be considered a "proletarian" nation, and should be ready to wage battle against European "plutocracies" such as Great Britain and France in order to get its fair share of world power. A small elite should set up a mass movement on that basis and propagate these ideas.[82]

As a matter of fact these and other initiatives helped the Italian government make up its mind to take over Libya in 1911. Gualtiero Castellini returned to Libya as a war correspondent and dedicated his book *Tuins e Tripoli* to his beloved uncle Scipio Sighele (they also made a joint trip there later). The intervention was received favorably by a wide spectrum of public opinion. After the vote in parliament, Corradini exulted: "Today everything is renewed, liberated, redeemed, purified. The people, the government, the mob, youth, the King, Rome and Italy are all redeemed."[83] The new mood in the country

82. On the Italian Nationalist Association see Vaussard (1961), Alff (1973), and De Grand (1978). Sighele recorded his experiences with irredentism and nationalism in three successive books: *Pagine Nazionaliste* (1910), *Il Nazionalismo e i Partiti Politici* (1911), and *Ultime Pagine Nazionaliste* (1912).
83. Quoted in Thayer, *Italy and the Great War,* p. 237.

even decided the government to further extend voting rights to some 8.5 million people – which amounted to near universal male suffrage. (Mosca, by the way, was one of the few to vote against it.)[84]

Sighele supported the Libyan intervention but opposed the authoritarian turn that the nationalist association itself was taking, and resigned. "Universal suffrage may not be the most perfect system," he wrote in an oft-quoted phrase, "but it sure is the least imperfect of electoral systems."[85] Meanwhile, his personal prestige continued to grow. After the Austrian government decided to expel him from the Trentino province, he was received by King Vittorio Emmanuele II, and met with Prime Minister Giolitti. In 1913 he suddenly fell ill and died. Large crowds followed the funeral procession in his beloved Nago, and a street was named after him in Milan. Although some other major crowd psychologists noted his early contributions to the field,[86] his academic fame was far outstripped by his reputation as a spokesman for the irredentist movement.

Because this chapter is the only one on Italy (compared to four on France), I was forced to extend it well beyond *La Folla Delinquente*, Sighele, and his immediate predecessors and successors. I cannot conclude, however, without following some of the trends discussed into their next phase. Both progressive liberalism and democratic socialism continued to lose momentum to new mass movements: revolutionary socialism and national chauvinism. After the First World War broke out, Italy at first remained neutral, but interventionists on both sides gradually got stronger. Mussolini and part of the left wing of the Socialist party broke away and founded a new daily, *Il Popolo d'Italia*. It was "either war or revolution," he said.[87] Corradini and the right wing of the Nationalist party fastened their hold on the weekly *L'Idea Nazionale* and transformed it into a daily, too. Both got substantial support from heavy industry in Italy, and some historians claim secret foreign funds were involved as well.[88]

Finally, the government decided to join the Entente powers in the war. But after the country had lost 650,000 people and spent 150 billion lire, its share in victory was disappointing. Italy was finally granted Trento, Trieste, Istria, and even Alto Adige, but many felt this

84. Michels even wrote that he was the only one, but other sources do not mention this.
85. *Il Nazionalismo e i Partiti Politici*, p. 173. Quoted in Garbari (1974), p. 542.
86. Tarde (1892 ff.) in French, Freud (1921) in German, and Park (1904, 1921) in English.
87. Thayer, *Italy and the Great War*, p. 280.
88. Alff, *Der Begriff Faschismus*, p. 71.

was not enough. Small "elites" tried to stir the great masses into further action. The writer and adventurer D'Annunzio led an expeditionary force that unilaterally seized the city of Fiume (making Pantaleoni his minister of finance, by the way, and helping Pareto to his long-desired divorce).[89] After there had been a sharp upsurge in revolutionary agitation (with socialists occupying factories and farms), Mussolini decided that the time had come to "transform a mob into a party."[90] He launched a series of fascist counterattacks, which culminated in the March on Rome, and his forced appointment as prime minister. Using precepts of both the crowd psychologists and the elite sociologists, he quickly expanded his mass movement as well as the totalitarian state.[91]

To this effect, he also needed a revised constitution and legal framework. A few years earlier, Ferri (who had long left the Socialist party) had headed a commission for a new penal law, to succeed the Zanardelli code, then considered outdated. This project was revived by the fascists, and adapted to the new situation. Many positivist ideas were retained, ranging from a strong emphasis on "social defense" and crime prevention, to a better adaptation to individual cases and concrete circumstances. One of the novelties introduced by the law was that it made an extenuating circumstance "having acted at the suggestion of a tumultuous crowd, except in cases of meetings or assemblies forbidden by law or by the authorities, when the guilty person is not an habitual or professional delinquent or contravener, or a delinquent by tendency."[92]

89. Tommissen, *Vilfredo Pareto*, p. 22.
90. As he told Emil Ludwig in a famous interview. Translation in Salomone (ed.), *Italy from the Risorgimento to Fascism*, p. 224.
91. As a former socialist, Mussolini was familiar with the ideas of the positivist school. More than the Italian crowd psychologist, however, French authors such as Le Bon and Sorel were his primary sources in this field. Similarly, the Italian-speaking Mosca held less prestige for him than the mostly French-speaking Pareto. Mussolini claimed to have followed some of Pareto's lectures in Lausanne, and later called his theory of elites the "most ingenious sociological conception of modern times" (Gregor 1969, p. 106). Furthermore, Mussolini met or corresponded with several of these authors. Mosca was appointed professor in Rome in 1923. But the reedition of the *Elementi* in that year (which had been prepared earlier, and which was also to become the basis for the English translation) scaled down the authoritarian and upgraded the liberal tendencies in the book. Pareto was made a senator the same year (and was even invited to become the Italian representative at the League of Nations in Geneva), but he declined because of health problems and died soon thereafter. Crowd psychologist Ferri and the elite sociologist Michels became close Mussolini allies – like many other ex-socialists and syndicalists.
92. Penal Code 1931, Book I, Part 3, Chapter 2, Article 62, Section 3. (British Foreign Office transl.).

Needless to say, this excuse for the actions of the mob took on an entirely different meaning in this specific context. The spontaneous rebellions of poor peasants that had originally inspired the rise of crowd psychology in the 1880s was replaced in the 1920s by the well-directed crowds of the *squadri fascisti*. Rather than being a challenge to the established order, "criminal crowds" now became its instrument.

3

A missing link: Fournial, anthropology, and the priority debate

Lyons 1892. Less than a year after Sighele's publication of *La Folla Delinquente*, another closely related monograph appeared, one that is rarely mentioned in the existing literature on the origins of crowd psychology. It was overlooked, for instance, by all continental authors – such as Boef, Moscovici, Mucchi-Faina, Rouvier, and Vlach. It was only mentioned in passing by Nye (1975), and somewhat more extensively by Barrows (1981), but without detailing any of the background as with the other crowd psychologists.[1] Yet, the monograph in question is a major missing link between the Italian Sighele's criminological and the Frenchman Le Bon's psychological approach.

Like Sighele's first book, it bore all the marks of a beginner's work: It was well documented, but poorly written. Like Sighele's book, Henri Fournial's *Essai sur la Psychologie des Foules – Considérations Médico-Judiciaires sur les Responsabilités Collectives* was primarily criminological in approach. But while Sighele and his mentor Ferri were lawyers, Fournial and his mentor Lacassagne were physicians. The emphasis in their theories, therefore, shifted further in the direction of medical psychopathology.

Whereas the Italian school of "criminal anthropology" had been led by Lombroso from Turin, the French school was led by Lacassagne from Lyons. Turin and Lyons are only a few hundred miles apart. In fact they are closer to each other than they are to their respective

1. The negligence is all the more surprising because Fournial's little-known book was mentioned in Tarde's better-known 1892 article on the subject.

capitals (especially after tunnels had pierced the Alps). At the time of the emergence of crowd psychology, the Turin and Lyons groups were involved in a heated dispute over the respective roles of biology and psychology in crime. The depth of the rift separating them is often overestimated, however. The Italians as well as the French were in the process of reorienting themselves, and were slowly moving from late neurophysiological to early psychodynamic approaches. The former were merely slower than the latter in giving up the one-sided search for organic correlates of deviant behavior. But before 1893 (the death of Charcot) the shift was far from complete in either country.

This is obvious from both Lacassagne's and Fournial's conceptualization of the mental mechanisms underlying crowd behavior. Fournial claimed, for instance:

> Under influences which we still know little of, something like a link between all the entities of a collectivity is created which unites them all, a kind of organization is formed in which each of the elements develops solidarity with the others: in that disparate assembly, a soul has arisen that guides this new organism. In sum, there is a very special being, to the highest degree capable of feeling and acting, but without intelligence; it is impossible to find in any of its actions the result of or even an attempt at abstraction, at induction and deduction. Reasoning is something completely alien to it, it is a spinal and ganglionic being – similar to those decapitated animals in which one observes an exaggeration of reflexes. The crowd, we finally say – using the expression of Prof. Lacassagne – is occipital, parietal; it is never frontal.[2]

Now, what does this mean exactly, and where do these ideas come from? In order to find the answer to this question, we have to take a closer look at the neurophysiology of this period, and at the physical anthropology to which it was closely related. It is noteworthy that anthropology was a field in which such diverse early crowd psychologists as Lombroso, Sergi, Lacassagne, Fournial, and even Le Bon were all active in one way or another. It turns out that they all shared certain preconceptions about psychophysiology, based on the supposed correspondence between mind and brain, brain and skull, head and body. Physical anthropology, therefore, was thought to hold the key to a wide range of phenomena: from cultural and social differences to deviant and criminal behavior.

2. Chapter I-1, p. 23 (my translation).

Colonialism, race, and physical anthropology

We have seen in an earlier chapter that the key to the ideological preoccupations of the historian Taine lay in the first seven years of the Third Republic: the French defeat against Prussia in 1870, the abortive Paris Commune revolt, and the subsequent protracted struggle over the nature of the new regime – culminating in President MacMahon's "legal coup" attempt of May 16, 1877, which backfired in the subsequent elections.

After that, it took another seven years before the republican framework had been entirely filled in. Grévy was made the new president and the parliament returned from Versailles to Paris. The Communards were granted amnesty and many socialists resurfaced; free, compulsory, lay education was introduced and political freedoms were extended. Finally, in 1884, the constitution was revised and universal (male) suffrage introduced. This completed the edifice of the Third Republic.

There was considerable social agitation during the early 1880s: wildcat strikes, riots, and bombings kept the Red Scare alive in certain quarters. Others, however, saw these as isolated incidents, and felt they hardly threatened the regime as such. Meanwhile, the regime also resumed a global foreign policy. The first Universal Exhibition in Paris had highlighted France's industrial take-off, and Prime Minister Jules Ferry advocated a corresponding colonial expansion, primarily to secure raw materials and access to markets but also to promote "l'égalité, la liberté, l'indépendance des races *inférieures*" (emphasis added).[3] Whereas Great Britain had by this time already subjected vast territories and populations, France was still largely limited to footholds, and Germany was lagging far behind. This situation is well illustrated by Table 1 on the expansion of the area (in millions of square miles) and population (in millions) of their respective colonial possessions.[4]

While France had acquired some colonies under previous regimes, it was only under the late nineteenth-century republican governments that these were extended and transformed into a real overseas empire embracing all continents.

Closest to home, in northwest Africa, France started to extend its

3. Speech in the Chamber of Deputies, July 28, 1885, quoted in Girardet, ed. (1966), p. 103.
4. Supan, *The Territorial Development of the European Colonies*, quoted by Lenin, among others in Wright, *The New Imperialism*, p. 30. Also see Zimmermann, *Der Imperialismus*, pp. 3–5, a.o.

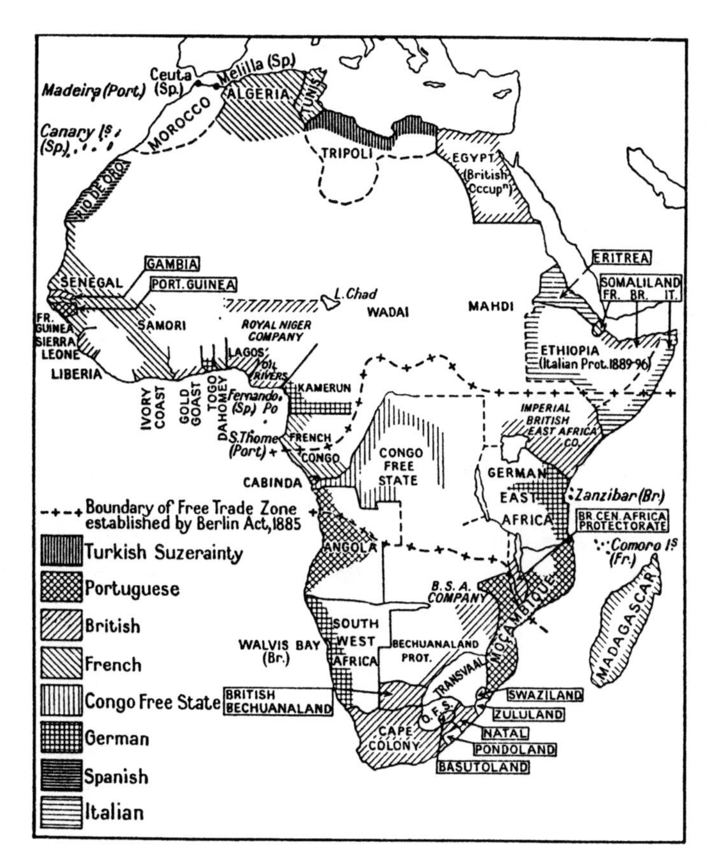

European influence in Africa, 1891. Note the French influence in West Africa, Madagascar, Tunisia, and Algeria, with Morocco still independent. Also note the Italian influence in East Africa, with Libya still under Turkish rule. (From Roland Oliver and J. D. Fage, *A short history of Africa*. Harmondsworth (U.K.): Penguin Books, 1962, Map 17. Copyright © Roland Oliver and J. D. Fage, 1962; reproduced by permission.)

hold from Algeria to Tunisia, and later to Morocco. Somewhat farther south, it started to extend its hold from Senegal and Guinea to a dozen other territories in western and equatorial Africa. In southern and eastern Africa it found the English firmly entrenched, but the huge island of Madagascar still "up for grabs." In northeast Africa France was not particularly lucky either, and was even forced to cede prime control over Egypt and the Suez Canal to the British. In southern Asia it retained only tiny footholds, but in southeast Asia it was able to extend its hold from the southernmost tip of Vietnam to the entire

Table 1. *Evolution of colonial possessions*

	Britain		France		Germany	
Year	area[a]	pop.[b]	area[a]	pop.[b]	area[a]	pop.[b]
1815–30	?	126.4	0.02	0.5	—	—
1860	2.5	145.1	0.2	3.2	—	—
1880	7.7	267.9	0.7	7.5	—	—
1899	9.3	309.0	3.7	56.4	1.0	14.7

[a]In million square miles. [b]In millions.
Source: H. M. Wright, *The New Imperialism – Analysis of Late Nineteenth Century Expansion* (Lexington, Mass.: Heath, 1961), p. 30

Indochinese peninsula. France furthermore acquired smaller possessions in the Indian and Pacific Oceans, and in the Caribbean – where it initiated the Panama Canal project. Thus, from the early eighties on, France quickly moved to become the second colonial power in the world.

Under previous regimes, the public justification for such conquests had been primarily religious zeal, but now secular goals such as economic growth, technical progress, and scientific discoveries became the official driving force. In the early 1870s, the Societé de Géographie de France had a mere 780 members. But by the early 1880s eleven other geographical societies had sprung up: Together they totaled no less than 9,500 members, one third of the membership of such societies around the world.[5] Ethnography and anthropology received a rapid boost as well. Under previous regimes laymen and military doctors had written extensive travel reports containing all kinds of intelligence on overseas countries and peoples. But institutional support had been marginal and intermittent at best. In the course of the 1870s, however, a complete infrastructure arose. A journal of anthropology was founded as well as a school and a museum.[6] The new network was closely related to military medicine on the one hand, and legal medicine on the other. For some time, anthropology became the central discipline among the emerging sciences of man and society, since physiological differences between individuals and collectivities were also felt to hold the key to psychological, social, and cultural distinc-

5. Brunschwig, *Mythes et Réalités de l'impérialisme colonial français 1871–1914*, quoted in Mayeur (1975), p. 125.
6. See Hammond (1980), Clark (1973), pp. 116 ff.

An engraving showing Zouave soldiers embarking for Tunisia in 1881. As soon as the regime of the Third Republic had been stabilized, France stepped up its efforts to expand its colonial empire. (From A. Castelot and A. Décaux, *Histoire de la France et des français au jour le jour*. Vol. 7: *1814–1902*. Paris: Perrin, 1977, p. 503.)

tions. And the most important physiological difference was taken to be race, of course.

By this time, the debate on race had been going on for a full century and a half. Linnaeus had not only classified plants and animals, he had

"The four races of man," as depicted in a textbook widely used in French primary schools in Fournial's day. The caption identifies the white race as "the most perfect." (From G. Bruno, *Le tour de France par deux enfants*. Paris: Bélin, 1877/1912.)

also divided humans into European, Asian, American, and African races. Count de Buffon, Cuvier, Saint Hilaire, and others had then worked out alternative classifications. This line of thought culminated in the work of Joseph Arthur Count de Gobineau (1816–82, Tocqueville's cabinet chief during his brief tenure as foreign affairs minister). His multivolume study *Essai sur l'Inégalité des Races Humaines* (1853–55) distinguished white, yellow, and black races, and argued that their physical and mental capacities differed substantially. Within the white race there were obvious contrasts between the Aryans and the Semites, he said, and among Aryans between Germans and French.

Not surprisingly, the latter theme became prominent after the Franco-Prussian war. One of the founding fathers of French anthropology, Jean Louis Armand de Quatrefages de Bréau (1810–92), immediately published a study on the "Prussian Race," which emphasized its barbaric traits. In several wholly or partly Germanic countries, teachers took to noting the observable feature of their pupils, in order to ascertain the racial origins of their respective peoples.[7] An anthropological handbook by Topinard in turn questioned the composition of the Gallic race, and suggested that it consisted of a mixture of the former elite (supposedly blond-haired and blue-eyed) and the subjected masses (supposedly black-haired and brown-eyed). Count Vacher de Lapouge even went one

7. In *Le Mythe Aryen*, Part 2, ch. 5, Poliakov claims that this research – inspired by Virchow – covered no less than fifteen million children in Austria, Switzerland, Germany, and Belgium.

step further by claiming that, during the French Revolution and its aftermath, the inferior groups had definitely taken over and were now phasing out the superior groups altogether. This heralded a call for racial purity in France, Germany, England, and other countries – we will encounter some of its consequences in later chapters.

Within the academic community the debate increasingly focused on the precise physical criteria by which races, nations, and even classes should be distinguished. Because the brain was increasingly considered the "organ" of the mind, the debate focused on cranial peculiarities. Under the Second Empire, the radical materialist determinism underlying this approach had run into powerful religious opposition, but under the Third Republic it now flourished freely. One of its key proponents was Paul Broca (Sainte Foy La Grande 1824 – Paris 1880). A professor of external pathology and later of clinical surgery in Paris, he became the founder of the first anthropological laboratory at the École des Hautes Études, and a co-founder of the first anthropological society.[8]

Like so many of his contemporaries, Broca had begun by emphasizing brain weight and skull volume. He claimed to have proven that the average brain was larger in the white race, and that it had continually grown in size over the course of European history. He also claimed to have proven that the average brain of eminent men was larger. It soon turned out, though, that the theory was not entirely consistent with the facts. Cuvier had had a rather large brain, for example, but Gall's had been just above normal (and so, as it turned out, was Broca's own). Gradually, therefore, an additional criterion was introduced: the form of the skull. It was expressed in a "cranial index": the ratio of the width and length of the skull. Superior groups had long and narrow (dolichocephalic) skulls, it was said, whereas inferior groups had short and broad (brachycephalic) ones.

After Darwin's theory *The Descent of Man* had been published, translated, and popularized in the course of the 1870s, the debate increasingly shifted to the purported similarities between inferior groups and higher animals, however. Facial angle was made a major criterion: A receding forehead and protruding jaws were supposed to be apelike. Certain facial traits were said to be apelike as well. Of course this was a delusion. Gould, who has reanalyzed and invalidated Broca's "hard" proof in this matter, noted: "The human body can be

8. In 1858 and 1859, respectively. The example was followed by London, Madrid, and Moscow in 1863.

measured in a thousand ways. Any investigator, convinced beforehand of a group's inferiority, can select a small set of measures to illustrate its greater affinity with apes. (This procedure, of course, would work equally well for white males, though no one made the attempt. White people, for example, have thin lips – a property shared with chimpanzees – while most black Africans have thicker, consequently more 'human' lips.)"[9]

In the process of trying to account for all kinds of newly emerging inconsistencies, the debate also shifted from exclusively external aspects of the brain, such as weight and form, to its supposed internal structure, such as the number of convolutions or the location of the brain stem (purported to indicate the relative proximity of the species to four-footed or two-footed animals). Neurophysiology had gone through a rapid development – ever since Magendie had distinguished the functions of the sensory and motor nerves, Flourens had distinguished the functions of cerebrum, cerebellum, and medulla oblongata, and Broca himself had even been able to locate a "speech-center" in the brain. In Broca's multivolume *Mémoires d'Anthropologie* (1871 ff.), his *Revue d'Anthropologie* (1872 ff.), and the École d'Anthropologie (1875), then, the debates increasingly centered on the relative development of the brain in superior and inferior races, in upper and lower classes, in males and females, and in adults and children.

After the spread of Haeckel's Law (that ontogeny recapitulates phylogeny), the latter difference was widely felt to hold the key: "Savages," workers, and women were often supposed to have infantile traits. Their cerebrum was less developed than the cerebellum and the medulla oblongata, and their frontal and anterior lobes were less developed than their parietal and middle lobes or their occipetal and posterior lobes, it was said. This became the central notion in Alexandre Lacassagne's theories, which in turn inspired Fournial's book on the crowd.

Criminal anthropology in France

In the previous chapter we saw that in Italy the criminal anthropology of Lombroso and his like initially focused on the supposed discovery of the "born criminal": an eternal recidivist who betrayed his atavistic

9. Gould, *The Mismeasure of Man*, p. 86.

nature by his apelike traits. In the preceding section we have seen that the physical anthropology of Broca and his like culminated in elaborate methods of corporal measurement. The famous statistician Adolphe Quételet (1796–1874) published a book on *Anthropométrie* in 1870. But another name, even an entire family, contributed significantly to the emergence of this field. In order to gain a correct understanding of the evolving divergence between the Italian and French schools of criminology during these years, we will have to make a brief excursion into the contributions of this family, the Bertillons.

The grandfather, Jean Baptiste, had been a demographer and a colleague of Quételet. His son Louis Adolphe (1821–83) was an anthropologist, a close collaborator of Broca in the Société and the École, and the head of the Paris bureau of vital statistics. He was succeeded there by his eldest grandson, Jacques (1851–1922), who developed uniform methods of recordkeeping and the "Bertillon classification" of the causes of death (which provided the first really comparable data for further analysis and thereby helped boost empirical sociology). The younger grandson Alphonse (Paris 1853 – Münsterlingen, Switzerland 1914) did not seem very promising at first, but in the end outdid them all by introducing the famous "Bertillonage method" of criminal identification.

He had been a rather mediocre student, but his father introduced him to the Paris Anthropological Society all the same, and probably stimulated his writing of a study on *Les Races Sauvages*. His father also introduced him to the Municipal Police Department as a clerk, and possibly helped inspire his original 1879 report, which proposed the systematic application of anthropometric methods to criminal identification. These new methods were particularly relevant for the "uncovering" of recidivists, who often gave false names upon renewed arrest, since a second offense was now considered an aggravating circumstance, usually leading to a heavier sentence.

After the practice of branding criminals had been abolished, the Paris Police Department had already taken to photographing them under the Second Empire, but for lack of adequate classification this did not solve the problem. After the Commune revolt tens of thousands of suspects were photographed, and for decades their pictures were used to highlight the "singular traits" of criminals and rebels. Bertillon, however, proposed an improvement of the physical description of criminals by meticulously measuring their corporal features. By measuring the length and width of their skulls (in order to identify

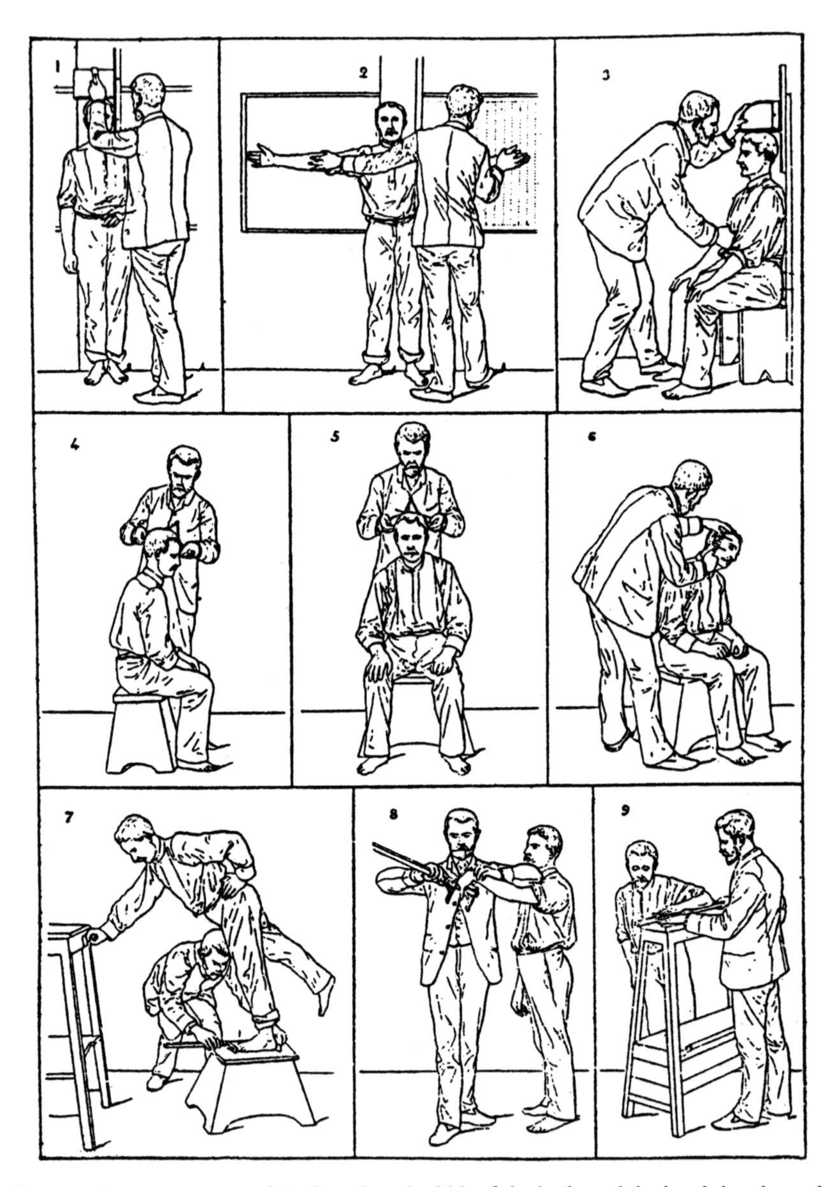

Systematic measurement of the length and width of the body and the head, hands, and feet as done by Bertillon and others. (From P. Vultus, *De physiognomiek van Lavater en de theoriën van Bertillon, Lombroso enz.* Amsterdam: Vennootschap Letteren en Kunst, n.d., Pl. 26, opp. p. 344.)

dolichocephalic and brachycephalic types), the length of the middle and little finger, and by classifying each feature as large, medium, or small, he obtained $(3)^4$ categories – that is 81 drawers in which to file cards in numerical order, and from which to retrieve them.[10]

His superiors were skeptical at first, but a probationary period in 1883 brought immediate results. The number of successful identifications of recidivists rose from 2 in the first quarter to 6 in the second, 15 in the third, and 26 in the last. The method was thereupon introduced on a larger scale, and Alphonse Bertillon quickly rose through the ranks. After having been made official director of a special Department of Judicial Identity five years later, he also proposed an improvement in photographic procedures: having both *en face* and *en profil* pictures taken, and by standardizing both pose and lighting conditions. By that time his fame had spread throughout Europe, and his "crime laboratory" had become a veritable tourist attraction for foreign criminological visitors.

At the same time, fingerprinting was slowly coming to the fore. In some countries, fingerprints had already been used for illiterates having to sign documents. Gradually it turned out that there were not only racial differences in the distribution of fingerprint types, but also that individual fingerprints as such were close to unique. An English medical missionary first "discovered" this in a Tokyo hospital, and wrote a letter to this effect to the journal *Nature,* and even (so he claimed) to the London and Paris police departments. In the late 1880s, several criminologists elaborated better methods of fingerprint identification. In Argentina (where the Italian school of criminology exerted considerable influence) a police official of Dalmatian extraction by the name of Vucetich was the first to publish a book on the subject, *Dactyloscopìa Comparada,* in 1888. In France Forgeot – a student of Lacassagne – wrote a thesis titled "Les Empreintes Digitales au Point de Vue Médico-Judiciare" (fingerprints from a medico-judiciary point of view) as early as 1891 and he also proposed a chemical method to make fingerprints fully visible. Lacassagne's close friend and colleague Bertillon contributed a study on professional anomalies in the fingerprints of seamstresses, laundresses, florists, and glassworkers. In England Conan Doyle, the creator of Sherlock Holmes and Watson (who had been an army surgeon himself) first referred to the technique that same year. Yet many consider Galton's *Fingerprints* (1892) the real breakthrough,

10. Rhodes, *Alphonse Bertillon,* Part 3.

since he claimed that there was only a chance of 1 in 64 million that two fingerprints were completely alike. In the course of the 1890s, then, anthropometry was gradually superseded again by fingerprints as the ideal method of criminal identification.[11]

The latter method proved superior, because it also operated on the "traces" that were often left at the scene of the crime (though hardened criminals soon learned to wear gloves, of course). We should add that both methods of criminal identification were introduced in certain colonies at an early stage. The white authorities there often felt that photographs did not solve their problems, because all natives "looked alike" anyway. In the major Indian province of Bengal, for instance, commissioner Sir William Herschel had made an early use of fingerprints on legal documents. Sir Edward Henry introduced anthropometric methods there as early as 1892, but started complementing them with fingerprints again five years later for purposes of criminal identification. Gradually, the technique was adopted in Europe as well. Another five years later, both the London and Paris police departments solved their first crimes on the basis of fingerprints alone. This sounded the decline of the Bertillonage method, and therefore the end of the elaborate measurement of criminal's heads.

The Bertillon episode, which is often omitted from studies of criminology in this period, is in fact crucial to a correct understanding of the evolution of the French school, and of its relations with the Italian school.[12] The establishment of the Italian *Archivio di Antropologia Criminale e Scienze Penali* (as it had been renamed) was followed six years later by the establishment of the parallel French *Archives de l'Anthropologie Criminelle et des Sciences Pénales.* But by the time Lombroso's school got around to propagating its "discovery" of the atavistic criminal with recognizable traits abroad, Lacassagne's school was just getting in the first really precise and large-scale results of corporal measurements of criminals. And although both Lacassagne and Bertillon remained inclined to admit organic correlates of social deviance, their data seemed to refute an all too simple link between external features and criminal inclinations. It was this circumstance that accelerated the French shift away from the atavism hypothesis and toward the degeneration approach – that is to say, away from physi-

11. Rhodes, *Alphonse Bertillon,* and Hibbert, *The Roots of Evil,* Part 5.
12. See, f.i., the otherwise illuminating study by Nye, *Crime, Madness and Politics in Modern France.*

ognomy and craniology and toward psychopathology and psychodynamics.[13]

Lombroso gradually altered his position somewhat, but he still stuck to the idea of a category of "born criminals" (see previous chapter). Lacassagne, on the other hand, was gradually converting to a belief in the primacy of environmental causes, invoking Quételet's assertion that society in itself "contains the germs of all future crimes."[14] "Societies have the criminals which they deserve," Lacassagne said. And he added, in a highly topical metaphor: "The social milieu is the mother culture of criminality; the microbe is the criminal, an element that gains significance only at the moment it finds the broth that made it ferment."[15]

When the Second International Congress of Criminal Anthropology was held in Paris in 1889, the Paris Anthropological Society was of course represented by a large delegation. Its members were gradually distancing themselves from their past errors, and were particularly keen on rubbing it in when discussing with the Italians. Broca's successor Topinard correctly remarked that the Italians did not properly distinguish between physical and social types. Topinard's colleague Manouvrier said Lombroso had not properly compared criminal and honest people by correcting for race, class, sex, and age.[16] The accusations produced a major row, after which it was decided that a mixed commission of seven experts would investigate the problem by comparing matched samples of one hundred criminal and one hundred "honest" people. That would settle the question.

Fournial and his "Psychologie des Foules"

Meanwhile, it would be incorrect to say (as some studies do) that the French had at this time already (around 1890) completely abandoned the idea of organic correlates of deviant behavior. They rejected the idea of atavistic features. But quite a few stuck to the idea of cerebral particularities that were sometimes thought to be reflected in head

13. Nye, *Crime, Madness and Politics,* ch. 4.
14. *La Physique Sociale* (1835). Quoted in Mannheim, *Comparative Criminology,* ch. 19 (p. 506 in the German translation).
15. *Actes du Premier Congrès* etc., pp. 166–7. Translated by Nye in *Crime, Madness and Politics,* p. 104.
16. *Actes du Deuxième Congrès* etc. The entire debate is well summarized in Nye, *Crime, Madness and Politics,* ch. 4.

form. This even holds for the leader of the French school himself, who was also the immediate source of inspiration for the crowd psychologist Henry Fournial.

This Alexandre Lacassagne (Cahors 1843 – Lyon 1924) had started out as an army doctor. He had been a student at the École de Santé Militaire de Strasbourg, and had been attached to the imperial army. After the Franco-Prussian war and the loss of the Alsace-Lorraine region, the school was transferred to Nancy, like so many other academic institutions. Lacassagne served briefly in Africa, where he was first confronted with both local "savages" as well as imported criminals (of the Foreign Legion).[17] After his return to France he worked for some time at the military hospital of Val de Grâce in Paris before embarking upon a civilian career as a forensic expert. He published minor studies on comparative criminology and major manuals on legal medicine. When a new Military Health and Legal Medicine School was set up in Lyons during the early 1880s, Lacassagne became a professor there. He soon founded an anthropological museum and a crime laboratory matching those in the capital.[18]

Most of his work focused on practical questions, but some of it had wider theoretical implications too. A few years before his appointment, for instance, he and a colleague published a study about "the influence of intellectual work on the volume and form of the head." Completely in line with the dominant ideas of those days, it claimed that intellectual work stimulated the development of the frontal lobe, and thereby affected the shape of the head. Two years after his appointment, he lectured on the Lombrosan theme of "criminal man compared to primitive man." Although it emphasized the role of the social environment, his definition of "social environment" was not as antithetical to "biological predisposition" as some present-day historians of criminology seem to think. As a matter of fact, he included biological predisposition in his definition of social environment:

> The social environment is an aggregation of individuals whose cerebral evolution is different. The superior layers, those which have evolved the most [in society that is] are the most intelligent: we may call them the frontal or anterior layers. The inferior ones, which are the most numerous, those in which the instincts pre-

17. This experience also contributed to his study on tattooing in savages and criminals, which refuted the Lombrosan thesis of their "atavistic" insensitivity to pain.
18. See Pinatel (1961), Souchon (1974), among others.

> dominate, will be the posterior or occipital layers. Between them, a series of layers, characterized by types in which impulsive behavior dominates, either inspired by instincts or by ideas: those are the parietal layers. Therefore, the march of civilization is necessarily very slow. Man is rather modifiable, but he improves only superficially. Before some progress is realized and an acquisition becomes organic, that is to say definitive, centuries and centuries are needed. When some influence, food, alcohol, education, an economic crisis, a revolution [!] affects a social environment, the preexisting cerebral equilibrium is upset and in the inevitable conflict produced, the anterior part of the brain is dominated by the posterior one, the phenomena of intelligence are dominated by instincts and acts. But if the cause is metaphysical thought, a philosophical speculation, a moral revolution, it never penetrates very far into the social layers, and the crowd remains insensitive to the seduction of the most scholarly theories.[19]

It should be added that Lacassagne stuck to such ideas throughout the 1880s, and still reconfirmed and applied them at the time of Fournial's thesis. In the preface to a study on habitual criminals in Parisian prisons, for instance, he again linked the same tripartition to both social class and criminal type. Frontal criminals, he said, had a disturbed thinking pattern: They were mostly alienated. Parietal criminals, by contrast, showed a disturbed behavioral pattern: They occasionally ceded to their impulses, and therefore threats and punishment might have some effect. Occipital criminals, however, had a disturbed emotional pattern: They were instinctual egotists, and there was little one could do about them.[20] These quotes show that the rift between Lacassagne's thought and that of the Lombrosans was not as deep as is often suggested. The typical hesitation in 1890 between the old physiological and the new psychological approach can easily be discerned from the dissertation of his pupil Fournial on crowd psychology, which is at the center of this chapter.

Henry Fournial (Trans en Provence 1866 – Paris 1932) came from the Mediterranean Var province. His father and uncles were timber merchants, they owned and operated sawmills and produced wooden

19. Lacassagne, *L'Homme Criminel . . .* , p. 7 (my translation).

20. See f.i. Lacassagne's preface in Laurent, *Les Habitués des Prisons de Paris,* p. v.

boxes in Le Muy near Draguignan. Rather entrepreneurial, they branched out into other activities, but (according to a present-day descendant) their fortunes fluctuated. Henry himself was twenty-three years old when he was drafted into the army. This would have broken off his studies, had he not been permitted to continue them at the École de Médecine Militaire in Lyon. He soon became fascinated with anthropology there, and avidly followed the lectures of Lacassagne. It was Lacassagne who suggested the subject of a thesis to him: a study of the criminal crowd like Sighele's, but with more of a focus on the underlying psychological mechanisms. The thesis was written in late 1891, presented on January 27, 1892, and jointly published by Storck in Lyon and Masson in Paris.[21]

Its starting point was Lacassagne's theory of the triunal brain, suggesting on the one hand that frontal types were usually underrepresented in crowds, whereas parietal and occipital types were overrepresented; and on the other hand that in the collective mind that emerged in crowds frontal functions were almost absent, whereas parietal and occipital functions dominated. Then, the book followed Lacassagne's classification of modifying factors.[22] Fournial first discussed the influence of *physical* factors on behavior: heat (climate, season); light (sun, clouds); meteorological phenomena (lunar position, atmospheric pressure); sound (noise, music); and movement. He continued by discussing the influence of *chemical* factors: air and water (lack of oxygen, excess of humidity); food and liquor (alcohol). He then discussed *biological* factors, related to individual differences in hereditary make-up, sex, and age. Finally, he discussed *sociological* factors, such as national and cultural particularities, but he claimed they affected individuals rather than crowds as such.

After having built this solidly material foundation, then, Fournial moved on to a more murky terrain: newly discovered psychological phenomena. Imitation, he said, was partly a normal social phenomenon, although bad instincts more easily led to "moral contagion" than noble sentiments. The religious and political history of both East and West showed that it was often the prestige of one man (or sometimes a woman) that was at the base of a wider social movement. The influence of this prestige was comparable to those discussed in recent

21. See the obituaries in *Le Matin*, September 5, 1932; and in *La France Militaire*, September 6, 1932.

22. Note that this approach is very similar to the one Lombroso chose for *Il Delitto Politico e le Rivoluzioni* in 1891.

publications on suggestion, hypnotism, and somnambulism. Occasionally, however, imitation could also take an even more morbid turn. This was well illustrated by the illusions and hallucinations that characterized epidemic madness. It was also highlighted in both fictional accounts and historical descriptions of the behavior of mobs, for instance during the French Revolution and the Commune revolt.

The conclusions to be drawn from this could be summarized in a dozen points, Fournial said. In order to demonstrate just to what extent he was a missing link between Sighele and Le Bon (and Tarde, for that matter), I quote them (almost) in full:

1. The crowd may be considered a being that feels and acts but that does not think. . . .
2. This individualization [emergence of a collective mind] is the result of the unconscious and suggestive diffusion of an emotion: It occurs through imitation or moral contagion.
3. Imitation can be explained by the instinct of vanity residing in man, and his need for approval of his acts; these sentiments make him copy what goes on around him, hoping in this way to gain the esteem of those whom he copies.
4. Imitation may proceed in a reciprocal or unilateral manner. When it proceeds in a reciprocal manner, its intensity is proportional to its subject, it varies with the number of elements [individuals], the degree of cohesion in the crowd, the nature and characteristics of the individuals that compose it.
5. When imitation proceeds in a unilateral way, it is comparable to the phenomenon of hypnotism: The intensity varies with the suggestive power of the cause, the point of departure of the imitation, and the degree of receptivity of the elements influenced.
6. Imitation may be normal, social, in which case it has the effect of propagating the sentiments, the acts . . . of ordinary life and produces: fashion, language, mores, etc., produces the great movements of enthusiasm, the great demonstrations.
7. Imitation may also be morbid: It acts on the troubled mental faculties, spreads illusions, erroneous sentiments due to a perverted imagination and bad instincts, producing the great neuroses, the great mental epidemics, crimes, and so forth.
8. The crowd may be occasionally criminal, and in this case

responsibility, without disappearing altogether, is considerably diminished, though in varying proportions.

9. The crowd may also be criminal by habit or may be pursuing its interest.
10. In both cases, it is not those who act who are solely guilty or most guilty but, whatever they actually do themselves, the leaders [*meneurs*], the agitators [*excitateurs*] are.
11. The responsibility of these latter ones is heightened, by contrast.
12. It is often dangerous to unite people in crowds. One can expect crowds to boost bad instincts rather than generous feelings. One may say that most often a "human beast" emerges from the crowd.[23]

Anyone familiar with the crowd psychology of this period will easily recognize that this summary by the unknown Fournial is a mixture of the ideas of the little known Sighele, the better known Tarde, and the most well known Le Bon. The crucial difference is that Sighele was quoted (albeit rather modestly), and Tarde as well (somewhat more extensively). Le Bon's book, however, bearing a shorter version of the same title, appeared more than three full years later, and mentioned neither Sighele nor Fournial. No wonder the Italian authors was outraged by the succession of French "piracies" of his book, and opened a fierce polemic – which even questioned the intermediary role of Lacassagne's close friend and colleague, Tarde.

Sighele's accusations are seldom taken seriously, even today. Rather than scattering the debates between the various protagonists over this entire study, I have therefore chosen to suspend the standard format of this chapter and devote a full section to the priority debate as such. I was able to trace various descendants or heirs of the founding fathers over the years, and to interview them. In some cases they still kept relevant correspondence from those days (see the list of interviews and letters under Unpublished Material in the bibliography). In some other cases, correspondence had already been published at the time, for instance in Ferri's journal *La Scuola Positiva*. By combining these elements, I have been able to go beyond the accounts by Nye and Barrows, and to show that – contrary to widely held belief – the young Italian author did have a fairly strong case, indeed.

23. Conclusions, pp. 107–9 (my translation).

Sighele, Tarde, Le Bon, and the priority debate

We have seen that around 1890, the debate on the hidden causes of mutual influence was gradually shifting ground. Concepts such as moral contagion, social imitation, and hypnotic suggestion were becoming more important. The first concept derived from a long tradition, but had recently been updated and applied to criminology by Aubry.[24] The second concept was central in the work of Le Bon, and we will return to its vicissitudes in the next chapter. The last concept was elaborated by Tarde, to whose development we will return in the final chapter. At this point, we will limit ourselves to noting that, as early as 1890, Tarde had published two books discussing the role of imitation in crime, which contained isolated pages on the crowd.[25] That year, the young Sighele even wrote to the older Tarde, asking for suggestions on the subject of *La Folla Delinquente*.

Tarde's descendants still hold most of the correspondence that he received during these decades. It permits a detailed reconstruction of the priority debate between the founding fathers of crowd psychology. In a letter from Sighele to Tarde, dated October 17, 1890, he wrote: "My first duty is to thank you for having been so kind as to react to my poor bibliographical notes – which did not deserve such an honour." After that, however, the roles were somewhat reversed.[26]

The further development of Tarde's own ideas received a strong impulse from the original edition of Sighele's book, which was published by Bocca in Turin in 1891. This can be inferred from the half

24. Apparently, this line of argument originated from a thesis by Prosper Lucas, entitled *De l'Imitation Contagieuse ou de la Propagation Sympathique des Névroses et des Monomanies* (1833). The theme was picked up again by Prosper Despine, who wrote a pamphlet against the current press reporting of crimes, *De la Contagion Morale – Faits Démontrant son Existence* (1870). Thereafter, P. Moreau de Tours fils published a larger essay, *De la Contagion du Suicide – À Propos de l'Épidémie Actuelle* (1875, 79 p.), which in turn inspired a thesis by his pupil P. Aubry: *De la Contagion du Meurtre – Étude d'Anthropologie Criminelle* (1887–88, 184 pp.) – which mentioned the crowd only in passing. Thereafter, Aubry also presented a paper on the subject to the Second International Congress of Criminal Anthropology in Paris in 1889. Later revised editions of his book (1894, 1896) contained an entire chapter titled "Les crimes des foules."
25. *La Philosophie Pénale* and *Les Lois de l'Imitation*. Note, however, that Tarde had not yet written a separate paper or article on the crowd. Curiously enough, I found during my trips that this misconception is even widespread among Italian historians of psychosocial science. Mucchi-Faina, for one, wrongly claims that Tarde's "Crimes des foules" was published in 1888, and thus entirely misses the point of Sighele's primacy.
26. Letter in the Tarde family archives, Paris.

dozen pages he devoted to it in a long review of several Italian works in Ribot's prestigious *Revue Philosophique*. They contained the nucleus of several of his later ideas, such as those on the similarities between crowds and sects (which in turn reinspired Sighele). The review was for the most part a matter-of-fact discussion, but it had a rather positive tone, and may have encouraged Alcan to publish a French translation of the Italian book that came out in mid-1892.

It would have been in line with current practice, if Sighele had presented a paper on the subject to the next (Third) International Congress of Criminal Anthropology, to be held in Brussels that same summer. But the Italians were still vexed by the accusations the French had made against their leader Lombroso during the previous (Second) Congress in Paris, and by the fact that the research commission set up to clarify the matter had never even gotten under way. Forty-six Italian criminologists wrote an irritated letter to the organizing committee, saying that under those circumstances, they saw no point in attending.[27]

This fact, and the publication of Fournial's book, may have contributed to Tarde's decision to elaborate his own ideas on the subject, and to present a paper titled *Les Crimes des Foules* himself. Just before he left for Belgium, on July 29, 1892, he wrote to Sighele:

> I have to thank you very much for having sent me the French translation of your *Folla Delinquente,* and also for having spoken about me in such sympathetic terms. At the same time I must apologize for having robbed you (*pillé*) by borrowing the subject of my report for the Brussels Congress from you. . . . In my written report I could only mention your brochure in passing, but at the Congress I will no doubt mention it more explicitly and extensively.[28]

Tarde's paper did indeed make a minor reference to Sighele's book. Barrows and historians of the episode tend to take for granted that he did indeed mention it more emphatically in his oral presentation.[29] The correspondence gives a different impression, however. Even worse: Someone (probably Ferri, who acted as the messenger of the Italians to Brussels) apparently told Sighele that *La Folla Delinquente* had not

27. Complete text in *La Scuola Positiva* 1892, pp. 423–4 (my translation). The decisive refutation of the Lombrosan thesis of the atavistic criminal came only in 1913 with the publication of Charles Goring's empirical study *The English Convict.*
28. Quoted by Sighele in Ferri et al., "Polemica sulla Psychologie des Foules," p. 372.
29. See Barrows's *Distorting Mirrors*, p. 157.

been mentioned *at all.* Upon hearing this, Sighele immediately protested to Tarde, and received the following reply (dated September 14, 1892): "In my [written] report you are mentioned as you should be. I have certainly been brief on your account. . . . I intended to be less concise in my oral report; but since I talked ex tempore . . . I was led into digressions that caused me to lose sight of your *Folla Delinquente.*"[30]

Verification reveals that Tarde had indeed been led into digressions – for instance when the discussion on the role of leaders was most appropriately interrupted by the arrival of the king in person. Tarde seized on the occasion to congratulate Belgium on the fact that joyful crowds were more common there than bloody disturbances.[31] But he also accepted rather exuberant praise of his "brilliant contribution" afterwards – without ever mentioning Sighele, it seems. Nevertheless, his frank admission closed the incident for the time being. Meanwhile both authors engaged in a cordial exchange of letters on the nature of the crowd, some of which were published in journals and included in later editions of the book.

Immediately after the Congress, reviews of the French translation of Sighele's book, *La Foule Criminelle,* began appearing in some of the most prestigious journals of the country. The neurophysiologist Charles Richet, for instance, wrote a short review in the *Revue Scientifique.* It was matter-of-fact, but positive in its overall judgment. The writer Victor Cherbuliez wrote a much longer review in the *Revue des Deux Mondes* (under the pseudonym G. Valbert). He proved rather skeptical about the positivist approach, but helped draw attention to the work all the same. The brief mention of the book in Tarde's paper, also published in Lacassagne's *Archives de l'Anthropologie Criminelle,* had the same effect, as did mention again in Tarde's follow-up article, "Foules et sectes du point de vue criminel," which was published in the *Revue des Deux Mondes* in 1893.

In sum, then, it was hardly possible for a French author interested in the subject to pretend to be unaware of the existence or even the tenor of the books by both Sighele and Fournial. Yet this is exactly what happened, when Le Bon (a friend of the editor of the *Revue Philosophique*) published two long articles in Richet's *Review Scientifique* in

30. Quoted by Sighele in the same "Polemica," p. 372.

31. He added that "big heads" were not always necessary to lead a crowd, and that in fact they were in fact often led by complete idiots. (*Archives* 1892, pp. 502–3).

April 1895. With some new material, they were soon published as a book by the same publishing house of Alcan (which Sighele then left in protest), under an abbreviated version of Fournial's title: *Psychologie des Foules* (1895). Sighele was outraged. It should be added that relations between the Italians and the French in general were not particularly warm for various other reasons at this precise moment.[32] But Le Bon's book really did it for Sighele.

He wrote a vitriolic review in Ferri's *La Scuola Positiva,* under the headline: "A Literary Piracy." He said it was a form of plagiarism that was the worst of its kind. "Do you want the proof? Here it is. And deny it if you can. The first chapter of your first book is a complete copy as to the line of thought and frequently a literal one as to its form. On pages 12 and 15 you summarize the introduction to my volume; on pages 17, 18, 19, 20, 21, 25, 26, 28, 30, 38, 39, 40, 45, 46, 47, you copy the ideas that I have developed in my first chapter."[33] According to Sighele, Le Bon also quoted exactly the same examples from the voluminous work of Taine, put the same emphasis on the role of *meneurs* and their "prestige," etcetera, etcetera. Therefore, Sighele announced that he would lodge a formal complaint for copyright violations with the "Societé des Auteurs" – both in France and in Italy.

At this point a key question arises. When did Le Bon get the idea for his *Psychologie des Foules?* A collection of Le Bon papers, which is now in the possession of a political admirer, mainly covering a much later period of his life, contains one item that seems to pertain to this question. It is a notebook that appears to contain successive scribbled outlines for his crowd psychology. The first outline is headed "La Psychologie des Foules et la genèse de leurs croyances" and has a date added to it, which may be ending with "91," although it is not entirely clear.[34] There is some reason for skepticism of this interpretation, though. First, both title and date were obviously added later, and were

32. Many Italian intellectuals such as Sighele had originally been francophiles, but then felt they were looked down upon by their French colleagues. Furthermore, military, political, and economic problems were developing between the two nations. Italy had realigned itself with Germany and even with its hereditary enemy Austria. Meanwhile, Italian workers fell victim to discrimination in Southern France, and dozens were even lynched in Aigues-Mortes in 1893. Similar incidents occurred in 1894. (See Romano, *Histoire de l'Italie*, etc., pp. 96–101.)
33. *La Scuola Positiva* 1895, pp. 171–3.
34. It may read 17-6-91, which was around the time Sighele's *La Folla Delinquente* was first published. The notebook is with the Le Bon papers kept by Mr. Pierre Duverger in Paris.

written in a different style, although apparently in the same handwriting. Second, the next outline is clearly dated "94," which suggests an exceptionally long interval. Third, the "first" outline is more elaborate than any of the five subsequent ones, whereas one would expect the reverse. But whatever one chooses to think of this "first" outline, it contains only a few of the elements that gave the final work its specific character.

It is possible, though, that Le Bon started toying with the idea of writing a crowd psychology as early as 1891, and even told others. In early 1892, after the publication of Sighele's and Fournial's books had made Tarde decide to produce a paper on the subject, he apparently wrote to their mutual friend Ribot to inquire whether Le Bon had worked on the subject. On March 30, Ribot replied: "Dear Sir, I am sure that Dr. G. Le Bon has never written any work on the psychology of crowds, and I doubt whether he ever will. If you want to correspond with him on the subject, however, here is his address: 29, Rue Vignon."

Ribot should have known, because 1892 was the year in which he and Le Bon founded a kind of debating club, the "Déjeuner des Vingt" (to which we will return). Two years later, and only six weeks after he had moved from Sarlat to Paris, Tarde was invited to attend one such meeting. In his diary he pokes fun at Le Bon's attempts to impress the ladies, but also proves flattered by the praise that he received for his own work. The two of them probably talked about crowd psychology, too. Tarde may have reminded Le Bon that he had published one or two articles on the subject, and may even have sent him a copy later. Because early in July 1894, Le Bon wrote a letter to Tarde, confirming "I have read your article, which I find quite remarkable."[35] As Tarde referred to both Fournial and Sighele, and as Le Bon was definitely embarking on the project, it is quite implausible that he did not read those two books.

Yet Le Bon's articles and book (published in 1895) only named Tarde in passing and did not mention Fournial or Sighele at all. The introduction to the first edition even explicitly claimed that the field was still "rather virgin land." After Sighele's protest, the second edition scaled this down to "rather unexplored," but the third edition reinstated the false claim that "there existed no general overview."

35. Letter from Ribot and letter from Le Bon both in the Tarde family archives, Paris (my translation).

However, it not only betrayed close similarities to Sighele's book, but to Fournial's as well. On the one hand, this can be seen in the comparison of specific elements, such as the physiological theories invoked. Just compare Fournial's theory, described earlier in this chapter, with the subtitles in Le Bon's first chapter. "The General Characteristics of Crowds": "The crowd is always dominated by the unconscious"; "Disappearance of cerebral life and predominance of medullary life"; "Lowering of intelligence and complete transformation of sentiments."

It can also be gathered from the general approach, and the psychological theories invoked. Le Bon claims that the transformation of the individual in the crowd leads to a "fixed orientation of ideas," whereas Fournial speaks of an "exaggerated generalization of idées forces." Le Bon says that this leads to "mental unity"; Fournial speaks of "unanimity of minds." Le Bon takes his prime examples not only from Napoleon's rule and the French Revolution, but also from the history of Christianity and Eastern religions; Fournial does the same. Although many of these ideas had already been quoted in the earlier literature, and their reinterpretation closely corresponded to the spirit of the times, these coincidences seem all too fortuitous.

As a better-known and more experienced writer, Le Bon shrugged off Sighele's accusations in the *Revue Scientifique*. "His little work is one of these honorable debutant's theses abounding in quotations but lacking a single personal view," he wrote.[36] Sighele responded with a high-handed letter to editor Richet, but the latter refused to print it. Ferri thereupon published the whole "Polemica sulla Psychologie des Foules" in *La Scuola Positiva*. In this polemic, Sighele also brought up his quarrel with Tarde in 1892 again. Tarde wrote a response blaming this change of mind on "occult influences" (probably a reference to Ferri's role in stirring up the animosity). But he added: "I am eager to put an end to this debate which I have not provoked and which – should it be prolonged – would unavoidably become slightly ridiculous."

Sighele then explained himself by publishing his correspondence with Tarde of three years earlier in *La Scuola Positiva,* and added that

36. *Revue Scientifique,* 1895, p. 635. Also see Le Bon's remarks in his *Psychologie du Socialisme* (1898), p. 108 and others – which twisted the meaning of *La Folla Delinquente,* however. See Ferri et al., "Polemica sulla Psychologie des Foules," in *La Scuola Positiva,* 1895, pp. 367–375.

his anger was not so much directed against Tarde, but rather against Fournial, and most of all against Le Bon. A few months later, on February 27, 1896, Sighele sent a private letter to Tarde, which is still in the family archives. It demanded detailed explanations on the attitude of some of the French involved in the dispute, but also proposed to restore peace in their own mutual relations. "I think that, after having explained oneself, one should close the matter and be greater friends than before." Tarde's answer was probably in a similar vein, and even gave Sighele the impression that he supported the accusations against Le Bon. This is apparent from another unpublished letter by Sighele, dated March 5, 1896. "Your last letter was so friendly that I do not know how to thank you for it," he wrote. "It is very comforting for me to see that you, at least, have not abandoned me! I do not need to tell you that all you have written to me about Mr. Le Bon will remain between you and me."[37]

Tarde did not express his reservations to Le Bon himself, however. Over the next few years, Le Bon repeatedly tried to win Tarde's esteem, but the latter kept a polite distance between them. This can be inferred from some of their mutual letters, which are still in the Tarde and Le Bon archives. After Tarde had published another major book in 1897, Le Bon wrote him a friendly though somewhat condescending letter on April 17, saying: "Bravo Chèr Ami, I am going to read your big big book, sure to find plenty of ideas there. . . . I am one of your great admirers." A year later, in May 1898, he confirmed that "At last, chèr ami, I have read one of your books. . . . I would be rather surprised if you had ever written a better one."

Apparently, Tarde did not show himself to be entirely pleased, because another Le Bon letter in June exclaimed that Tarde was "ticklish" and made an oblique reference to attacks to which they had both been subjected. A month later, in July, Le Bon thanked Tarde for quoting him: "the first time that my name slips from your pen . . . you had become the only sociologist today ignoring me altogether" (sic). Le Bon must then have had two of his books sent to Tarde: probably his new book on the psychology of socialism, and also his earlier book on the psychology of primitive and civilized peoples. A letter from Tarde to Le Bon, dated March 13, 1899, states that the latter's work "on French colonization" was entirely in line with his own ideas. "I therefore have to thank you for having provided a

37. Both letters are in the Tarde family archives in Paris (my translation).

confirmation [of my theories] – all the more precious because they seem to have been totally unfamiliar to you" (sic).

A new letter from Le Bon to Tarde, another eight months later, and dated November 26, suddenly came up with the charge that Sighele had now "plundered" the work of Le Bon. Le Bon tried to enlist Tarde for a revanche: "Between compatriots one must have a short memory" [with regard to differences], but "with foreigners – and above all with foreigners who hate us as profoundly as the Italians do – one must, I believe, have a very long one. . . ."[38] Apparently, Tarde turned down the invitation for a joint punitive expedition, although it was obvious that Sighele had indeed plundered Tarde, at least in *La Delinquenza Settaria* and subsequent works. During these same years, Tarde kept at a polite distance from Sighele as well, and Sighele's attempts to enlist his support for lectures in Paris were not entirely successful, it seems.

Although Sighele was probably right in his original claim at primacy, then, this did not prevent Le Bon from getting all the credit – both from contemporaries and subsequent authors on the subject. Le Bon's *Psychologie des Foules* (which was indeed more mature and better written than its predecessors) became an international bestseller, and Le Bon came to be widely considered *the* founding father of crowd psychology par excellence. This in sharp contrast, by the way, to his subsequent claims – related by Nye – of being the true inventor of Nietzsche's theory of eternal return, and even of Einstein's theory of relativity.

The mysterious silence of Fournial

In the row that developed in late 1895, early 1896, one figure was conspicuously absent: Fournial. Although his contribution to the emergence of crowd psychology was modest, one cannot help wondering why he did not respond to Sighele's harsh accusations. In the same letter in which Sighele resolved his differences with Tarde, dated February 27, 1896, Sighele reported that he had already sent three letters to Fournial's publisher Storck "to get my due," but had never received a reply. A week later he wrote that he had just fired off a fourth letter, "and should I not receive an answer within a week, I will write to Mr. Lacassagne." None of the histories of this whole episode contains any clue as to the reasons for this persistent silence, but a personal file on

38. Letters from Tarde kept in the Le Bon papers kept by Pierre Duverger in Paris. Letters from Le Bon in the Tarde family archives in Paris (my translation).

Fournial in the Val de Grâce Military Hospital casts some light on the question.

The reason his publisher Storck had trouble getting in touch with Fournial, and why Fournial did not bother to respond, was simply that in those years he was an army doctor attached to colonial expeditions in the jungles and deserts of Africa; he had entirely different things on his mind than his academic prestige. First he was assigned to a large force dispatched to conquer the island of Madagascar for France. But in early 1895, already 60 percent of the troops had been hospitalized with tropical diseases. By early 1896, more than 30 percent of the fifteen thousand men were dead, and Fournial himself returned to France a convalescent.[39]

But he remained there only a short while. In the early summer of 1896, he was attached to the Foreign Legion, and late in the summer he was sent back to Africa, where later he joined Fernand Foureau and François Joseph Lamy's famous expedition to cross the Sahara from Algeria, and to converge on Lake Tchad with two other expeditions, setting out from Senegal and Congo, respectively. This time, 20 percent of the men died, and Fournial also fell ill again.[40] This explains why he did not bother to claim a stake in the early development of crowd psychology. As a matter of fact, a promotion report in his personal file in the military hospital of Val de Grâce in Paris carries the express mention "none" under the heading "scientific work."

Yet, Fournial *was* involved in some other scientific work, since the habitual role of a military doctor on such colonial expeditions was to collect samples of plants and animals, to measure and depict the native population, etcetera. Some of his drawings were even included in subsequent books on the expeditions.[41] Here again is an obvious parallel with Le Bon, who was also a physician interested in anthropology and in expeditions reconnoitering prospective colonies.

It is more than a mere coincidence, therefore, that in the work of both the crowd is frequently likened to savages, whose cerebral faculties were supposedly less developed and whose instincts and reflexes were supposed to be more dominant anyway. In this respect it is interesting to mention one last event, though it might seem beyond the

39. See, f.i., Gén. Reibell: *Le Calvaire* [!] *du Madagascar.*

40. See, f.i., Cdt. Reibell, *Le Commandant Lamy.* Also see *Hommage à Fernand Foureau.* Fournial's role in this expedition is mentioned in the articles by Méd. Gén. Jean des Cilleuls, a.o.

41. Gén. Reibell, *L'Épopée Saharienne.*

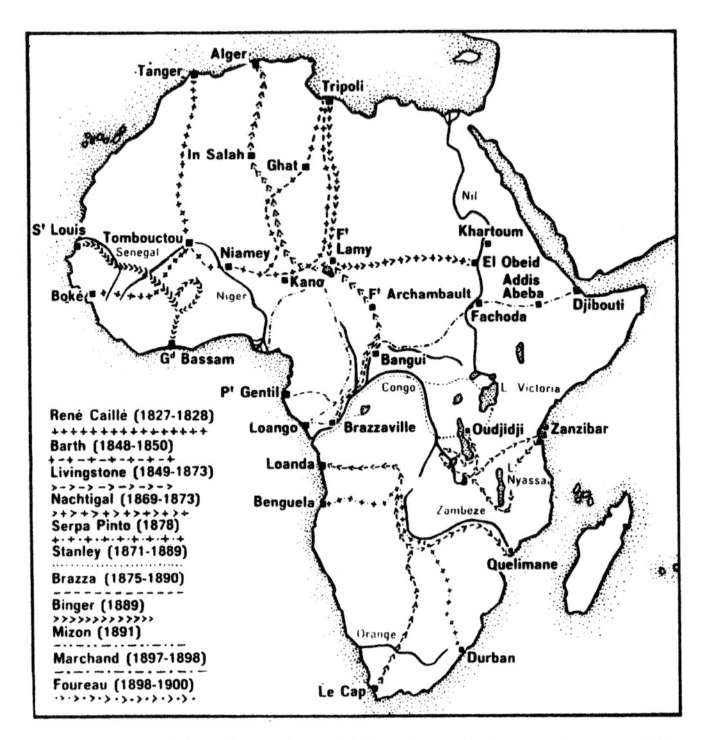

Major European expeditions in Africa, 1827–1900. The last major expedition was led by Foureau; Fournial was the expedition's biologist and doctor. (From C. Grimberg and R. Svanström, *Histoire universelle*, Vol. 11: *De la Belle Époque á la Première Guerre mondiale*. Fr. ed. Verviers: Marabout, 1965, p. 76.)

scope of this study. This concerns one of the major exploits of Fournial's career: the confrontation with a hostile mob of "primitive" Arabs, some fifteen years later.

In the same year the Italians invaded Libya, the French disembarked in Morocco in order to "persuade" the sultan to accept a protectorate. In spring 1912, the departure of an official mission from Fez was delayed by a violent thunderstorm, however. A military newspaper reported: "towards noon of the 17th of April a gang of violently overexcited armed Moroccan soldiers storms out of the Chérarda Kashbah – shouting threats of killing the French. Other soldiers, as well as common people and escaped prisoners join this band of madmen, convinced that the hour of the long desired and secretly aspired Holy War has finally arrived."[42]

42. *La Caducée – Colonies, Guerre, Marine,* August 3, 1912 (see also October 5).

Another source continued:

> Shooting breaks out in town. Cries of the "imams" and "youyous" of the women sound in answer to the shouts of the mutineers. The lowest rabble of Fez joins the rebels. A massacre is started, commissioned and noncommissioned French officers are killed in the streets or at their homes. . . . The hospital immediately becomes the nucleus and the stronghold of the resistance. . . . They had to hold out for a full four hours. It is here that we find the energetic physician-major Fournial again. As soon as the riots started, he immediately had the Auvert hospital put into a state of defense. He resolutely took command of whoever he could find: the musicians of the band of the first riflemen, the orderlies, the sick, and even the wounded who were able to stand on their feet – he immediately armed them all. He spread his troops, assured the guards of gates and walls; through his sang froid and ardor he managed to electrify everyone and to communicate his flame.[43]

Fournial succeeded in holding out until reinforcements arrived, although he was wounded a few months later, supposedly by a "hallucinating" Algerian. He was mentioned in parliament for his bravery, decorated, and promoted to chief physician for the entire Fez region, where a few years later a street was even named after him. Toward the end of the First World War, he was made medical inspector of the eastern armies, reportedly imposed a "sanitary dictatorship" on a large part of the Balkans in order to prevent epidemics from spreading, and was promoted to general.

Later, he retired to Paris, remarried with a much younger woman, and had the family finance his attempt to become a right-wing member of parliament. His grandnephew still remembers a lively electoral meeting in Draguignan, his charismatic personality and imposing look – but the bid failed. His personal papers were later destroyed. He died in 1932. His grave on the Le Muy cemetery carries the proud mention of no less than "twenty-three decorations." But the marble is crumbling, and half grown over with *laurier rose*. And he has long been forgotten as the fourth founding father of the "Roman" (or "Latin") school of early crowd psychologists.

43. Théveney in *La France Militaire*, September 14, 1932 (my translation).

4

The era of the crowd: Le Bon, psychopathology, and suggestion

Paris 1895 was the setting for the publication of Le Bon's *Psychologie des Foules*. On the one hand, it relied heavily on previous authors but went to great lengths to dissimulate and minimize this debt. On the other hand, Le Bon's book was indeed a much better written and habile synthesis, which put more emphasis on mass movements in general, and appealed more directly to the sensibilities of the middle class. Gordon Allport, in his authoritative history of social psychology, even went so far as to plainly call it "perhaps the most influential book ever written" in that field.[1]

Le Bon's book is not only the most popular of the early crowd psychologies, its background is also among the most discussed of all the texts we are dealing with. Since Nye's 1975 landmark study on Le Bon, authors like Apfelbaum, Barrows, Geiger, McGuire, Métraux, Moscovici, Rouvier, Thiec, Vlach, and others have produced major articles and books devoted wholly or in large part to the book. Although sometimes ignoring similar elements in his predecessors, most of them agree that the psychological models underlying the book were twofold: It borrowed on the one hand heavily from the evolution/dissolution theory of the so-called Paris School, and on the other hand from the hypnotic suggestion debate between the Salpêtrière and Nancy schools. We will return to the meaning of these terms later.

"It will be remarked," Le Bon said, "that among the special characteristics of crowds there are several – such as impulsiveness, irritability, incapacity to reason, the absence of judgment and of the

1. See Allport (1954), p. 26.

critical spirit, the exaggeration of the sentiments, and others besides – which are almost always observed in beings belonging to inferior forms of evolution – women, savages and children, for instance."

One of the reasons for this seemed to be

> that an individual immerged for some length of time in a crowd in action soon finds himself – either in consequence of the magnetic influence given out by the crowd, or from some other cause of which we are ignorant – in a special state, which much resembles the state of fascination in which the hypnotized individual finds himself in the hands of the hypnotizer. The activity of the brain being paralyzed in the case of the hypnotized subject, the latter becomes the slave of all the unconscious activities of his spinal cord, which the hypnotizer directs at will. The conscious personality has entirely vanished; will and discernment are lost. All feelings and thoughts are bent in the direction determined by the hypnotizer. . . . The conclusion to be drawn from what precedes is, that the crowd is always intellectually inferior to the isolated individual, but that, from the point of view of feelings and of the acts these feelings provoke, the crowd may, according to circumstances, be better or worse than the individual. All depends on the nature of the suggestion to which the crowd is exposed.[2]

Various studies of Le Bon also pointed to the political background of the book: the "crisis of mass democracy in the Third Republic" (so Nye); and in particular, the succession of the Boulanger episode, the breakthrough of the workers movement, and the Panama scandal. Some have even pointed to the Dreyfus affair, although that had really only begun to unfold. Yet various authors limit themselves to a mere mention or description of these circumstances, and thus miss the point. This point, to which others do allude, is that these events were really the first expressions of larger trends, which came to haunt Europe for the next half century and more. Like the French Revolution and the Commune revolt, the Paris agitation of 1885–95 (and beyond) had an exemplary character.

At the outset of that period, France had become the first major European nation with universal (male) suffrage. The Boulanger episode between 1886 and 1889 can be shown to have been the first major step in the collective learning process leading from a Bonapartist to a protofascist formula of mass mobilization: one able to attract support

2. Le Bon, *The Crowd* (1895a), pp. 35–6, 31, 33, respectively.

from both the Left and Right. The rise of the workers' movement, furthermore, was not just a smooth continuum, as some authors would have it, but it made a huge leap between 1889 and 1893–95 in France. The Panama scandal, which reached its peak in 1893, not only shattered public confidence in parliamentary democracy but also coincided with the anarchist terror campaign of 1892–94, which culminated in the killing of the French president. Only when the political background is presented in this way does the true meaning of the *Psychologie des Foules* fully reveal itself.

The rise of these mass movements, Le Bon said substantially, poses a fatal threat to elite liberalism. If we do not want the masses to succumb to the socialist spell, we will have to reinforce the nationalist alternative by providing strong leadership, he felt. This is the key to the uninterrupted sequence of four bestsellers that Le Bon began to write in 1893: his psychology of peoples (1894), of crowds (1895), of socialism (1898), and of education (1902). The message of these works is summarized well in one sentence from *Psychologie des Foules* quoted above: "All" (read: the survival of France, and of Western civilization in general) "depends on the nature of the suggestion to which the crowd is exposed. . . ."

In the rest of this chapter, we will examine these books and their backgrounds in the same way we dealt with the previous major works on crowd psychology. Since Le Bon was already well over fifty when he published these bestsellers, one section is devoted to Le Bon's earlier life and work. The second section delves into his intellectual background, and into the two new psychopathological paradigms from which he borrowed heavily. The third section analyzes the social background, and the new types of mass movements that came to the fore during these years. Only the fourth section goes into some more detail on Le Bon's book on the crowd, and on the three other related books of that same period. Finally, the last section deals with Le Bon's later life and work, and with his considerable impact on twentieth-century political thought.

Le Bon as a medical popularizer and a colonial anthropologist

Gustave Le Bon descended from a provincial family of civil servants and military men. He was born in 1841 in Nogent-le-Routrou near Chartres, attended a lycée in Tours and medical school in the Paris of the 1860s, at the time of the triumph of positivism.

He completed his studies in 1866. He also claimed he obtained a doctoral degree at the German University of Heidelberg the subsequent year, but secondary authors found no evidence to support his claim. He then started a private practice and began writing and traveling. During the lost Franco-Prussian War and the abortive Commune revolt of 1870–71, he headed a military corps of volunteer ambulances near the capital. His strong pronationalist and antisocialist inclinations were further reinforced by these events. When peace returned, he resumed civilian life and scientific work, even producing some minor inventions. He also resumed work as a medical popularizer: Over the years, he wrote studies on such diverse subjects as alcohol and tobacco, scientific photography and horse training, premature burial and human reproduction. Several of these books, and particularly the latter, were relatively successful and made him reasonably independent financially.

One of his most wide-ranging early works was *La Vie,* officially published in 1872 (though it may actually have come out a bit later). Nominally, it was a book about human physiology, but it showed that he was well-acquainted with the latest discoveries of contemporary psychology as well, including the earliest stirrings of the evolution-dissolution theory and the hypnotic suggestion debate.[3] It even included various references to their possible bearing on mass phenomena, such as military exercises (p. 577), religious sects (p. 581), and collective hallucinations (p. 582).

As a general rule, Le Bon closely monitored scientific innovations and often took up new approaches before they had caught on, trying to improve and apply them himself. He saw himself as a dauntless pioneer, and ended up claiming priority in a wide range of subjects, not only crowd psychology but other areas as well. This fits the description of Le Bon by others as a rather vain man. He also seems to have been an ardent womanizer. His one long liaison was with a married author, who reportedly incorporated his ideas, his personality, and even their relationship in her novels.[4] But neither he nor his brother had any

3. One forerunner to whom Le Bon apparently owed a lot was Alfred Maury (1817–92), remembered today as a sleep theorist, but also keenly interested in myth making, nation building, racial questions, and other topics. His *Le Sommeil et les Rêves* was first published in 1861, but reached its fourth (revised and extended) edition by 1878. It identified a hierarchy of mental functions, and their successive appearance/disappearance upon waking up/sleeping in. Its tenth chapter, on ecstasy, had several points in common with Le Bon's later views. Nye (1975, p. 26, 37n.) found an extensive summary of the book in Le Bon's handwriting. Also see Ellenberger (1970), pp. 303 ff.

4. The novelist in question was a notorious antifeminist, writing under the pen name of Jeanne Loiseau. She was the wife of Daniel Le Sueur. See Nye (1975), p. 85.

A French military infirmary similar to that in which Le Bon worked during the war of 1870. (From A. Castiglioni, *Storia della medicina*. Milan: Mondadori, 1936, p. 747.)

children – a fact that contributed to the strange itinerary of his personal archives after his death.[5]

I have already alluded to the fact that Le Bon was linked in various ways to the scientific milieu of his predecessors. He had been an army physician like Lombroso, Lacassagne, Fournial, and others. It was precisely out of military medicine that legal medicine and even criminal anthropology arose in these years. We have seen that many of his predecessors were active in those fields. Furthermore, they were closely related to physical anthropology and race studies, to which Le Bon turned in the late seventies. He was even a member of the Paris Anthropological Society (mentioned in the preceding chapter), participated in its meetings, and published in its journal. He wrote various articles comparing the skull size and shape of different races, sexes,

5. Nye (1975, pp. 16–17, n. 2) says that the estate was divided. The vast bulk of his personal papers came into the possession of the Carnot family. But Le Bon's secretary Antoinette Clotten kept "a sizeable selection of those materials which she presumably felt would be most useful in the building of his historical reputation." Today, the former part is in the hands of Pierre Duverger, a political admirer and the man behind "Les Amis de Gustave Le Bon" in Paris, which reedited a number of his books (1894, 1898, 1910, 1912, 1913, 1924) and brochures.

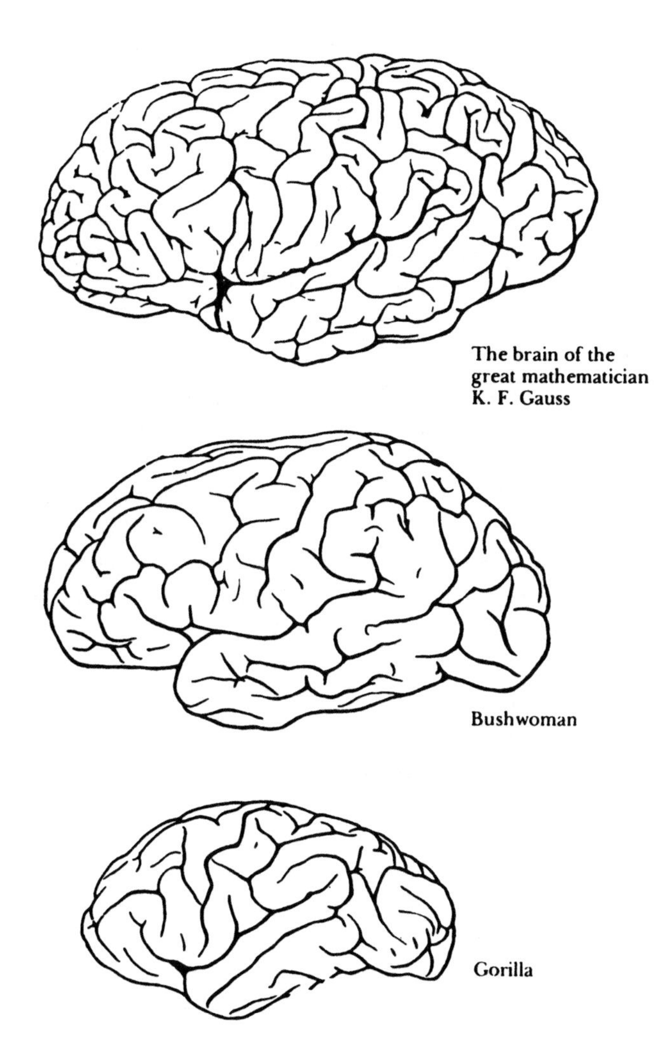

A comparison of the brain size of supposedly more- and less-developed primates. This type of study was at one point done by Broca and also by Le Bon, but later abandoned. (From S. J. Gould, *The mismeasure of man*. Harmondsworth, U.K.: Penguin, 1981, p. 90. Also in E. A. Spitzka, A Study of the Brain of Six Eminent Scientists and Scholars, etc. *Transactions of the American Philosophical Society*, 21:175–308.)

classes, and "great men" (some of which Lombroso quoted in his book on revolutions).

In the course of the eighties, however, the dominant thought was gradually moving away from a strictly physiological emphasis on race toward a more psychological emphasis on nationhood. Taine had originally spoken of "la race, le milieu et le moment" as the determining factors of individual and collective life. But in *Les Origines de 'la France Contemporaine* he had devoted ample room to mentalities as well. Taine's friend and colleague Renan, a noted historian of civilization, changed his mind too. In 1882 he gave a famous lecture at the Sorbonne on the question "What is a nation?" The answer was "A nation is a soul, a spiritual principle . . . [it is based upon] the common possession of a rich legacy of memories from the past and consent in the present, the desire to live together and the will to continue to develop one's heritage."[6]

The republican regime was trying to foster precisely such memories and desires in national education. One well-known example of such an undertaking was a new geography book used in the (now compulsory) primary schools. *Le Tour de la France par Deux Enfants – Devoir et Patrie* (a tour of France by two children – duty and fatherland) was a loving depiction of various regions, illustrated with some two hundred extremely evocative engravings. By 1887 (within ten years of its first publication) some three million copies had been circulated and had impregnated a whole generation with a nationalist fervor.[7] The schoolbook had been written by someone with the pen name G. Bruno: a woman who was to remarry the philosopher Fouillée. It may be more than a mere coincidence, then, that this same Alfred Fouillée (1832–1912) first developed the theory of "ideo-dynamism" between 1886 and 1889. Elaborating this in his *L'Évolutionisme des Idées Forces* (1890) and *La Psychologie des Idées Forces* (1893), he claimed that rather than being a passive reflection of social reality, certain ideas may become an active force in shaping it. He later applied these notions to the psychology of the French people, and of European peoples in general. This same theory strongly influenced Le Bon, Sorel, and others in their emphasis on the importance of political myths.

Throughout the 1880s Le Bon was involved in colonial anthropology. It started with another attempt at academic respectability, a grand synthesis in Spencerian and Darwinian (or rather, Lamarckian) style: *L'Homme et les Sociétés* (1881). Its two volumes, more than fifty chapters and almost one thousand pages, presented a comprehensive

6. French in Girardet (1966, pp. 65 ff.), English in Zeldin (1979–81, Vol. III, pp. 13 ff.).
7. De Pater (1987); Mayeur (1973), p. 115.

overview of contemporary knowledge: the beginning of the universe, matter and force, the origins of life, race and civilization, the evolution of man and society, the role of great men and great creeds, institutions and the family, education and language, property and law, religion and morality, and finally the rise of modern industry. His thesis was that men are born unequal, that they can improve themselves only through a constant struggle for existence, that cultures rise and fall as a result of the superior dominating the inferior or vice versa – in a biological, as well as a psychological and a sociological sense. Here again, Le Bon proved to be well-informed about novel theories. The work already contained the first nuclei of many of his later ideas, albeit in a very elementary form.

The book was meant to be the opening work of a ten-volume history of oriental civilizations, which was to direct colonial expansion in the Near, Middle, and Far East. The very next year, Le Bon applied for an official grant to travel to the Arab world, to study its race and culture. When the request was denied for lack of funds, he undertook the journey at his own expense. This resulted in an immense study, *La Civilisation des Arabes* (1883–84). Thereafter, he applied for an official grant to travel to the Indian subcontinent. This time his application, supported by an influential relative, was accepted.[8] The trip resulted in another large study on *La Civilisation de l'Inde* (1886), which was well received by some colleagues, but criticized by other scientists for its sloppy methods of quotation and its excessive claims to originality.[9] This may have shattered his hopes for academic acceptance, and further fostered his popularizing tendencies. In a subsequent work, *Les Premières Civilisations* (1889), he tried to go one step further. He discussed the rise and fall of the various great civilizations of the Middle East, and tried to deduce the general laws governing them.

Obviously, he was already thinking about applying them to contemporary France and other European nations. Like its predecessors, however, it counted some eight hundred pages and was hardly suitable for the "instruction" of a larger public. Therefore, Le Bon decided to

8. Primarily Sadi Carnot (1837–94), a distant relative with whom he was in touch regularly. Carnot was elected vice president of the Chamber in April 1885 and became a minister of commerce and finance shortly thereafter. Two years later he was elected president, under circumstances to which we will return. The later editions of Le Bon's India book were dedicated to Carnot, who was thanked for his "*haut patronage.*"

9. Nye (1975), ch. 3.

redirect his energies to writing one or two smaller books, in a clear, direct, and apodictic style, addressing the grave dilemmas of his time. In the end, this resulted in four successive books on the psychology of peoples, crowds, socialism, and education – to which we will return later. But first, we will take a closer look at the ten years immediately preceding their publication. For it was in these very years that the psychological paradigms with which he felt affinity were greatly refined, and the historic trends that worried him became more pronounced.

The Paris School of psychopathology and the Salpêtrière–Nancy debate

Le Bon's two most popular books were those on the psychology of peoples and of crowds; both were closely related and published in 1894–95. The former book was dedicated to his friend Richet, who had published ideas on hypnotic suggestion from the midseventies on, and figured prominently in the Salpêtrière–Nancy debate. The latter book was dedicated to his friend Ribot, who published ideas on mental evolution and dissolution from the early seventies on, and is considered the founder of the so-called Paris School. It is not surprising, therefore, that Le Bon hinted at these notions throughout his work. However, it was only between the mideighties and the midnineties that the Salpêtrière–Nancy debate *and* the Paris School really emerged on a larger scale. Le Bon followed the tide: His psychology of peoples and crowds is largely based on an adroit combination of those two models: of the hypnotic suggestion and of the evolution–dissolution model.

Both models arose out of abnormal psychology, but were applied in normal psychology, too. It has been pointed out that the major contributions to early psychology in France and Austria derived from case studies in psychopathology, whereas the major contributions in Germany and America derived from laboratory experiments in psychophysiology. It is not entirely clear whether this was a mere coincidence, or somehow related to different national traditions and styles. Whichever the case, this contrast should not be overestimated. Both Taine and Ribot drew attention to the importance of unconscious phenomena as well as the need for a systematic study of mental functions.

Just like Taine, Théodule Ribot (1839–1916) was well-acquainted with scientific developments in neighboring countries. His first major work was *La Psychologie Anglaise Contemporaine* (1870–71), which discussed the work of the Mills, Spencer, Bain, and others. His subsequent dissertation on psychological heredity claimed that a constant environment led to habitual behavior, which in turn crystallized into innate reflexes after some time.[10] It also included a discussion on the role of willpower in political and military leaders, and the formation of national character (ch. 7 and 8, resp.). Later, Ribot translated Spencer's psychology, together with Espinas. The latter's doctoral dissertation explored the biology and psychology of "animal societies" and included several fragments on the social (and crowd) psychology of animals – which were in turn quoted by several of the social and crowd psychologists analyzed here.

Ribot went on to incorporate many of Spencer's ideas on evolution and dissolution, particularly their seminal elaboration by the neurologist Jackson.[11] In the course of evolution, he said, higher species have gradually acquired higher mental functions. These higher functions are less developed in primitive humans, and more developed in civilized humans. Mental disturbance consists of a dissolution of higher functions, and a resurgence of lower functions. Similar intuitions had reportedly been formulated by earlier French authors too.[12]

Another major work of Ribot, *La Psychologie Allemande Contemporaine* (1879), discussed the work of Herbart, Fechner, Wundt, and others. He also wrote a book on the philosophy of Schopenhauer. (A French translation of fragments from Nietzsche only appeared in the early nineties.) Meanwhile, Ribot founded the *Revue Philosophique*, which soon became a major forum for the psychological debates of these early years.

During the first half of the 1880s, Ribot set out to develop his own system based on the careful study of mental disturbances. According to Bernard, the founding father of experimental physiology in France, diseases were the "experiments" (*expériences*) of nature. Therefore, Ribot made three successive studies: on the diseases of memory, the diseases of will, and the diseases of personality. His conclusion was

10. Rather than Darwinian, this theory might therefore be labeled Lamarckian. Note that these various cultural theories from Le Bon to Freud often have a strong Lamarckian streak.
11. Alfred Espinas (1844–1922) is particularly relevant in relation to the elaboration of notions of gregariousness. I have described some relevant developments on this subject in a conference paper on Wilfred Trotter (Van Ginneken 1987), and a preliminary chapter meant to be included in a sequel to this study. John Hughlings Jackson (1835–1911), author of *Brain and Mind* (1859), editor of the neurological journal *Brain* (1879 ff.) most fully expressed these ideas in an 1884 lecture on "Evolution and dissolution of the nervous system."
12. Ellenberger (1970) mentions Moreau de Tours as a case in point.

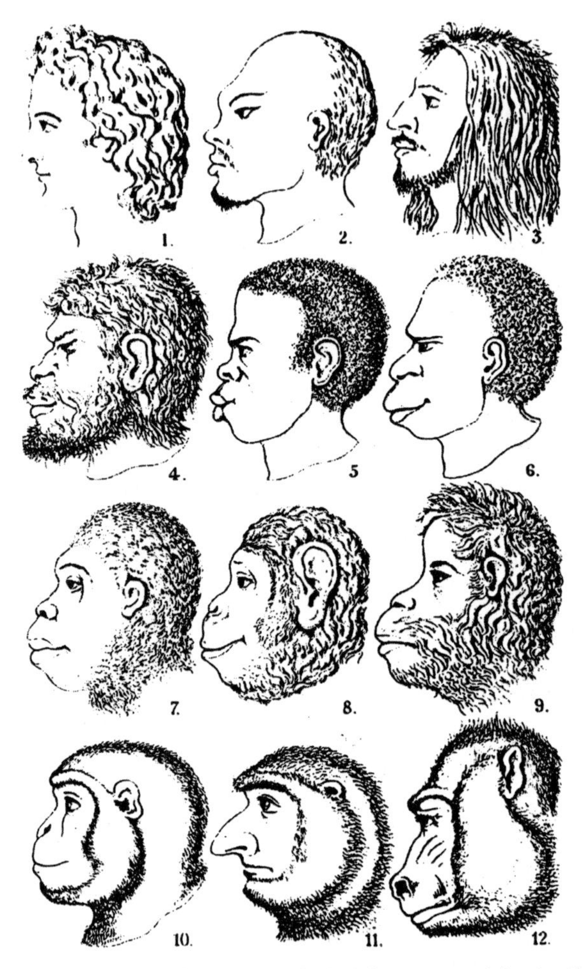

Heads of different kinds of primate in ranked order, as stated in the first book on human evolution by Ernst Haeckel, published in 1868. (From R. Boakes, *From Darwin to behaviorism*. Cambridge: Cambridge Univ. Press, 1984, p. 213. Original by Wellcome Trustees, London.)

that they were all based on some form of loss of mental control. "Le moi est une coordination," he said in a famous phrase: a capacity to integrate and synthesize various mental functions.[13] Ribot then published various articles resulting in *Psychologie de l'Attention* (1889), which discussed normal functioning and mental disturbances, ranging

13. *Les Maladies de la Personnalité*, Conclusion ("The I is a coordination").

from an incapacity to concentrate to its opposite, "idées fixes," or fixed ideas.

Around the midnineties, Ribot published more articles, resulting in his seminal *Psychologie des Sentiments* (1896), which discussed normal functioning and mental disturbances, ranging from mood swings to excessive passions, "the idées fixes among feelings." This book is one of the major keys to Le Bon's work on peoples and crowds. Among others, it discussed the "folie du pouvoir," occurring when a psychological predisposition coincides with a lack of external restraint; and the "délire des grandeurs," occurring when there is an excessive need to manifest oneself in the outside world (pp. 249–50). Finally, it concluded that dissolution often led to a form of regression, to an "infantilisme psychologique": manifesting itself in extreme changeability, impulsiveness, and lack of willpower (p. 422).

From 1892 on, Ribot and Le Bon were constantly in touch, as they co-chaired a monthly dinner and debating club. It is not surprising, then, that Le Bon's psychology of peoples and of crowds should use exactly the same model. Civilized men, Le Bon said, have a more evolved mind than savages, women, and children. But under certain circumstances, everyone's higher functions (conscious will, rational thinking) may dissolve, and lower functions (emotional feelings, unconscious reflexes) take over. This either may be the result of a long-term process, such as the weakening of national character, or it may be the result of a short-term process, such as the derailments of agitated crowds. But most often, he felt, these two processes were closely related.

Ribot's theory of psychic evolution and dissolution was gradually picked up and elaborated by younger scholars as well. On the one hand, this included Alfred Binet (1857–1911), the second director of the first psychological laboratory in France (at the Sorbonne), and the founder of the first French periodical in the discipline (*L'Année Psychologique*), famous for his subsequent "invention" of the intelligence test. During the late eighties and early nineties, he published several studies on mental (dys)functioning, such as *Les Altérations de la Personnalité*. On the other hand, there was Pierre Janet (1859–1947), the successor of Ribot at the Collège de France, and well known as the major proponent of psychodynamic theory in the country. In the late eighties and early nineties, he, too, published several studies on mental (dys)functioning, such as *L'Automatisme Psychologique*. Their ideas

"The Hypnotist" by Daumier. (From A. A. Roback and T. Kiernan, *Pictorial history of psychiatry and psychology*. London: Vision Press, 1969, p. 61.)

on personality change and psychological automatism also found an echo in Le Bon's crowd psychology.[14]

Much of their research during this period focused on the empirical study of one particular condition under which some kind of psychic dissolution seemed to take place: hypnotic suggestion. These activities were in turn related to a second major development of these same years: the Salpêtrière-Nancy debate.

Ever since the days of Mesmer, there had been an ongoing discussion on mesmerism, animal magnetism, somnambulism, hypnotic suggestion, hallucinations, and other forms of altered consciousness – whether induced by another person or occurring spontaneously. These phenomena fascinated laypeople, as is well illustrated by their prominent place in the French realist and naturalist literature of the nineteenth century. But they fascinated a small minority of medical people, psychiatrists, and early psychologists, too. Although the reality of the phenomena was denied again and again by "official" science, every single decade produced its own slate of new studies vindicating them.

14. On Janet and psychodynamic theory, see Ellenberger (1970, ch. 6). On Binet, Janet, and crowd psychology, see: Nye (1975), Barrows (1981), Moscovici (1981), Apfelbaum and McGuire in Graumann and Moscovici, eds. (1986a), and McGuire (1986c).

Thus, the 1860s – with their triumphant emergence of positivist science – also produced a number of important studies on these controversial subjects by luminaries such as Broca and others.[15]

Around the midseventies, the phenomenon of "somnambulism" drew the attention of Charles Richet (1850–1935), a student of the great Bernard himself (the founder of experimental medicine in France), who was about to head the *Revue Scientifique,* and later would become a physiology professor at the Ecole de Médecine (and even win a Nobel Prize). He reported that in one of his various experiments with hypnosis, he had been able to make a Bonapartist woman cheer a radical leader – a true miracle![16] According to Ellenberger's monumental study, it was this same Richet who first pointed out the reality of these phenomena to Charcot, whose scientific star was rapidly rising in the small world of Paris.

At the time, Jean Marie Charcot (1835–93) was only a professor of anatomical pathology, but he soon was to occupy the first chair of neurology as well. He worked at the famous Salpêtrière hospital, which had become a true "village within a city," housing some five thousand patients, mostly women, in "about 45 buildings with streets, squares, gardens, and an old and beautiful church," according to Ellenberger (p. 93). Charcot's main interest was in a special ward for convulsionary cases – mostly epileptics and hysterics. He soon found that the hysterical symptoms could be evoked or revoked under hypnosis. From the late seventies on, he went public with this discovery.

His assistants usually "rehearsed" the performance with the patients, the preferred one, Blanche, being an actress herself. Then Charcot, nicknamed the "Napoleon of the Neuroses," "demonstrated" them to the audience assembled in a lecture hall: not only students and physicians, but also lawyers and journalists. Dress, presentation, lighting: Everything was done to enhance the dramatic effect.[17] Le Bon attended these meetings on various occasions, and according to his biographer Nye (1957, p. 30), it was here that he met several of the protagonists of the hypnotic suggestion debate. Charcot (who, it seems, never hypnotized anyone himself) maintained that hypnosis worked only with hysterics, and that hysteria was the result of an

15. Azam, Lasègue, Liébault, Maury, etc. See Barrucand, *Histoire de l'Hypnose en France.*
16. Reported in *L'Homme et l'Intelligence,* p. 541. Note that such a confusion of political identities was exactly the feat realized by Boulanger a few years later, and commented on by Le Bon.
17. One such scene is depicted in a well-known painting by Brouillet. See Ellenberger (1970) between pp. 330 and 331.

organic predisposition. The debate emerged in the early 1880s, when this "Salpêtrière" position was challenged by the Nancy school.

Ten years earlier, the cession by France to Germany of the Alsace-Lorraine region (which was not only the major area of physical, but also of spiritual contact between the two nations) had led to the transfer of the French University of Strasbourg to Nancy. Near Nancy, a medical doctor by the name of Liébault had long been practicing hypnotic suggestion as a curative therapy free of charge, and had even written a book about it. In the early 1880s, he was already treating more than a thousand patients per year. The fact was brought to the attention of a medical professor at the university, who went there to investigate and gradually came to concede that it did indeed work.

This professor, Hippolyte Bernheim (1840–1919), found that his therapy not only worked with hysterical women, but also with normal men – albeit in differing degrees. It was not based on an organic predisposition, he felt, but on mental mechanisms. Soon, Bernheim started publishing books opposing Charcot's views and elaborating his own. His first book on suggestion (1884) described how hypnosis accentuated ordinary phenomena such as credulity, obedience, and excitability (ch. 7). But in his third book, *Hypnotisme, Suggestion et Psychothérapie* (1891), he offered an even more complete theory. He defined suggestion as "the act by which an idea is introduced into the brain and accepted by it" (p. 24). Furthermore, he added, "Any idea suggested and accepted tends to transform itself into an act, that is to say a sensation, image, movement" (p. 31). Normally, this "cerebral automatism" is "moderated by the superior faculties of the brain, the attention, the judgment which constitute cerebral control" (p. 49). But when these faculties are weakened, automatism is reinforced. This is what happens in sleep, hypnosis, and suggestion (p. 80). And it is this mechanism that leads to illusions and hallucinations (p. 116). It is a purely mental mechanism, Bernheim concluded, which has little to do with organic predisposition.

From the early 1880s on, then, there was an escalating debate on the true nature of hypnotic suggestion and related phenomena: with the powerful Salpêtrière school in the capital on the one hand, and the emerging Nancy school in the province on the other. The fad was soon picked up by the larger public. Hillman (1965) quotes a contemporary article from *La Médecine Populaire,* for instance, which observed that everyone was talking about magnetism "in drawing rooms, journals and reviews. It is practiced in hospitals, scientific societies,

Table 2. *Evolution of publications on hypnotism according to M. Dessoir*

Year	Number
1880	14
1881	39
1882	39
1883	40
1884	78
1885	71
1886	131
1887	205

theatres and homes" (p. 166). In the late eighties, Dessoir compiled the first bibliography of some eight hundred studies on these subjects, which had been published in the course of the decade by almost five hundred authors in fourteen different languages (60 percent being in French). It revealed a rapid increase, as shown in Table 2.

Over the first four months of 1888 some 71 studies were published, which pointed to a possible annual peak of 213. By that time various specialized journals and societies had sprung up as well.

But the real explosion came only in 1889, the year of the Universal Exhibition in Paris, which attracted unprecedented crowds (some thirty-three million visitors). Many foreign scholars came to the French capital to participate in one or more of the dozens of scientific congresses organized on that same occasion. Among them, at least four included sessions and papers on hypnotic suggestion: the international congresses on magnetism, hypnotism, physiological psychology, and criminal anthropology. Their participants included most founders of psychology (including most of the French mentioned here), and also most of the founders of crowd psychology, ranging from Sighele via Tarde to Freud. It is probable that Le Bon attended at least some of these sessions. But whatever the case, he was already well-acquainted with all sides of the debate, as can easily be discerned from his books.

Whereas the Salpêtrière school retained the upper hand in the debate during its expansion between 1886 and 1889, it gradually lost out to the Nancy school between 1890 and 1893, when Charcot and his followers were forced to admit that some of their results were based on artifice. Curiously enough, however, Charcot's power was so great that

Collective hysteria at the sound of a gong, as depicted in a French study of 1883. (From H. Vogt, *Das Bild des Kranken*. Munich: Lehmanns Verlag, 1969, p. 314. Original in P. Richer, *Études cliniques sur la grande hystérie ou hystéro-épilepsie*. Paris: Delahaye, 1883.)

people like Binet, Janet, and Richet long maintained that they supported his position, whereas a close reading of their texts reveals that they began leaning toward Bernheim instead. It was only after Charcot's death in 1893 that his reputation and that of his school eroded rapidly, whereas that of Bernheim and his school grew accordingly. By this time, a large part of the psychological pioneers had accepted that everyone's "I" consisted of various competing levels, ranging from largely conscious to completely unconscious ones. It is not surprising, then, that Le Bon incorporated these notions in his psychology of crowds (in the same year that Freud (and Breuer) incorporated them in their epoch-making *Studies in Hysteria*).

Although there was a gradual shift from physiological to psychological explanations, one should remember that many of these authors were convinced materialists. Several of them turned to invisible forces as a major explanatory device. At an earlier stage, mental contagion had often been likened to some kind of infection. After that, chemical emissions and heat exchange had become important metaphors. Now magnetism and electricity came to the fore, soon to be replaced by forms of radiation and other mechanisms. We have seen that the possible role of such factors in crowd behavior had already been discussed by Fournial. But a further elaboration was undertaken by Jean Luys, another close friend of Le Bon.

In 1894, at a time when Le Bon himself was working on the *Psy-*

chologie des Foules, Luys published another laudatory review of Sighele's *La Foule Criminelle* in the specialized *Annales de Psychiatrie et d'Hypnologie.* "These phenomena of centrifugal radiation, these caloric, electrical, magnetic, neural forces which form a part of living beings," he said, "make contact with similar surrounding individuals. There is a fusion, a sympathetic or antipathetic impregnation of living beings which in this way touch each other at a distance through fluïdum irradiation. . . . they fuse with one another through the unconsciously perceived incitations of neuro-magnetic streams. . . . all these radiant energies result in a very intense electro-magnetic potential, with enormous dynamic power." During theatrical performances and mass meetings, he added, this manifested itself in bursts of applause or sudden cries: "Long Live the King" or "Long Live the League."[18] McGuire has shown that such ideas on invisible radiation also played a major role in the psychical research of those days.[19]

From the very start, then, the hypnotic suggestion debate was closely linked to collective behavior patterns. Mesmer and his followers already favored the simultaneous treatment of larger groups.[20]

The founding father of French psychiatry, Esquirol, had pointed to the existence of psychic epidemics in a major handbook.[21] Le Bon's friend Richet did the same.[22] And so did the great Charcot.[23] We should also bear in mind that all these phenomena were discussed simultaneously in Italy and other countries, and that forerunners of these theories had already been incorporated in the crowd psychologies of Sergi, Sighele, and others.

One of Charcot's close associates was Paul Règnard (1850–1927), at the time deputy director of the physiological laboratory at the Sorbonne, and a co-author of the famous *Iconographie de la Salpêtrière,* a multivolume book of illustrations of mental illness. From 1881 on, Règnard gave a series of lectures for the Association Scientifique de France that dealt with animal magnetism, and also with dancing manias, demoniacal possession, drug fads (!), and "delusion of grandeur." In 1886 these lectures were published under the collective title *Les Maladies Épidémiques de l'Esprit* (the epidemic illnesses of the mind).

18. Luys, Études de Psychologie Sociale, pp. 294–5 (my translation). On Luys's friendship with Le Bon see Nye (1975).
19. The Collective Subconscious: Psychical Research in French Psychology, 1880–1920.
20. Ellenberger (1970), pp. 63–4.
21. *Des Maladies Mentales,* pp. 29, 501, 586, etc.
22. *L'Homme et l'Intelligence,* pp. 548 ff.
23. *Oeuvres Complètes,* Vol. I, p. 335; Vol. III, p. 235; etc.

The introduction discussed the excesses of mass meetings and crowd phenomena in general (p. xi). The epilogue discussed the question of the possible explosions of the next century: the moral dissociation of the human race, the excessive egoism of the ordinary [!] man, the threatening destruction of present society. "I fear," he concluded, "that the epidemic diseases of the mind in the twentieth century will be the folly of slaughter, the madness of blood and destruction."[24]

In his 1891 book, Bernheim, too, concluded that suggestion dominated the history of mankind, its wars and revolutions, its alternate movements of generosity or hate. "The people become an angel or a devil because they are suggestible."[25] Bernheim stressed that this applied to everybody, although in different degrees. Others emphasized that "inferior" people were most vulnerable. Another Charcot associate published a book on demoniacal possession in Central Africa. The great man himself edited a series of reprints called "The Diabolical Library."[26] In his various works, he pointed to religious ecstasy and dancing manias in European history, and to an older German monograph by Hecker on psychic epidemics.[27] Ellenberger and other authors have demonstrated that exorcism was a major predecessor of hypnotism. Bernheim and Charcot, both major spokesmen for secular science in the Third Republic, even went so far as to link miraculous cures during the Lourdes pilgrimages to hypnotic suggestion.[28]

There was a general agreement among various scientists that the building of the great religions, cultures, and nations had begun with mass movements inspired by strong leaders, often somewhat neurotic themselves.[29] Nationalism was a modern faith, they said, and so was socialism. "Men of the people, docile minds, former military men, craftsmen, subjects used to passive obedience" were more suggestible,

24. P. 429 (my translation).
25. *Hypnotisme, Suggestion, Psychothérapie*, ch. 11 (my translation).
26. See Ellenberger (1970), pp. 6, 95, respectively.
27. In the first volume of his collected works (p. 336), he referred to Hecker's study on *Die Tanzwuth* (1832, dedicated to Humboldt). It covered dancing manias (in which people danced themselves into a trance) of previous centuries in Germany, Holland, and Italy, but also pointed out more recent examples from Africa, Western Europe, and North America. In its conclusion (p. 63), it mentioned imitation and sympathy as a possible explanation. Hecker also wrote about plague epidemics (1828, 1832) and their mental aspects.
28. Bernheim (1886), p. 218; (1891), p. 51; Charcot (1892) quoted by Guillain (1955), pp. 177 ff.
29. See, e.g., Letourneau's famous *Physiologie des Passions*, pp. 224 ff. Note that he was a major figure in the Paris Anthropological Society, in which Le Bon was active for some time, too.

Bernheim said, than "refined spirits" (1884, p. 6). Therefore: all depended on the nature of the suggestion to which the masses were exposed.

Boulangism, socialism, and antiparliamentarianism

Some of Le Bon's background and the origins of crowd psychology have already been covered well by the existing literature, for instance in the aforementioned works by Nye and Barrows. This chapter inevitably duplicates some of this earlier work, but in this section I will further explore the mass movements dominating the ten years or so preceding the publication of the *Psychologie des Foules* and related bestsellers.

In order to do so, we will have to keep in mind that crowd phenomena had changed very much in both form and content over the preceding decades, as is highlighted in successive specialized studies by Hobsbawm, Rudé, the Tillys, and others. France was the first major European country to introduce universal (male) suffrage in 1884, and this suddenly made the political problem of the relation between elites and masses, and between organization and spontaneity much more acute. The Right was faced with the problem that religious and traditional justifications for monarchical and aristocratic rule were rapidly losing their appeal. It had to find and develop new themes, which might attract a new majority of voters. The Left was similarly faced with the problem that violent protest had become increasingly counterproductive. It had to develop new forms of peaceful propaganda and agitation in order to channel popular support into a viable movement. Extremists on both sides split off, and sometimes found common ground in their opposition to the bourgeois republic. Permanent political mass organizations gradually became a common reality. They differed not only in their slogans and symbolism, but also in the specific combination of techniques they employed for mobilizing supporters and intimidating opponents. Some capitalized on financial support, others on volunteer groups. Most opened offices, started journals, introduced membership, held meetings. New methods of communication and transport were employed. This collective learning process spanned many years and even decades, covered entire countries and even continents. Out of it, certain basic formulas arose – more or less standard configurations of political theory and practice that came to characterize the major currents and regimes of the next century.

Throughout this development, certain times and places were "laboratories." From the mid-1880s on, Paris was just such a major battleground, as was reflected in the general debate on the nature of the newly founded mass democracy.

Most critics were struck by the obvious contradiction between the behind-the-scenes maneuvering of the financial and intellectual bourgeoisie on the one hand, and their public mass rhetoric aimed at pleasing "the people" and "the crowd" on the other. A most remarkable and relevant text in this regard is Raoul Frary's *Manuel du Démagogue,* which appeared as early as 1884.[30] Just as there are no sovereigns without courtiers, he said among other things, there is no democracy without demagogues (p. 4). "The motives that touch the crowd are in no way mysterious; in order to get to know them and use them, some general principles and notions of history suffice" (p. 8). Don't be too clear too quickly: "By the discretion of your language and the vagueness of your imagery, you will seduce everyone, and you will get the crowd on your side" (p. 24). "We are not any more in love with money and power than other peoples, but we are more eager to please the master, to march with the crowd. . . . Throughout the ages they [the French] have sought the smile of the prince and the applause of the people; today prince and people are one" (p. 36). And: "there are a hundred good pupils of Machiavelli for one good disciple of Montesquieu" (p. 86). The book reads like a premonition of things to come.

Meanwhile, politicians, too, questioned the real functioning of democracy. One early critic was the journalist and senator Edmond Schérer (1815–89), a Protestant theologian of Swiss origin (a smaller country which had had a longer experience in this field). He pointed to the contrast between "le pays actif" and "le pays passif," and to Ferry's distinction between "a kind of political *elite*" (emphasis in the original) and "the great working masses, who busy themselves about politics only on election days. . . . The elite of which the president of the council speaks is made up of professional politicians. They are men of liberal or semiliberal professions." The key to the functioning of the official political institutions was the electoral committee, he said. "It must [however] not be imagined that the electoral committee is composed of members regularly nominated, and still less that it expresses

30. Quoted from the 1981 partial reedition *Du Bon Usage de la Mauvaise Foi.* Since there are no page numbers, I have counted the first so-called French title page as page 1 (my transl.).

the feelings of the people, whose organ it professes to be. More frequently it is self-constituted, and always finally composed of the ringleaders of every locality" (1883/1884, pp. 22–3). "The people are often led, in fact we might say always," he continued, "but the leaders obtain the confidence of the masses only by yielding to their instincts, and conforming to their habit of thought and desires. They guide rather than follow the mob. They foster the passions that it is in their interest to excite; or if they excite passions, it is only by fostering them" (p. 43).[31] Similar reflections were published by politicians in partly francophone neighboring countries, such as Switzerland and Belgium.[32]

Although Napoleon had created a stable administrative structure, governments under the Third Republic succeeded each other at an alarming rate. In seventy years, it was to have no less than some fifty prime ministers, almost ninety cabinets, and four to five hundred ministers.[33] There was an ongoing debate on possible revisions of the republican constitution, pitting various currents against one another. Some of the radical group, for instance, favored a reinforcement of the power of the president and the executive. The former deputy and later senator Alfred Nacquet (1834–1916) published several major articles criticizing "Parliamentarianism" and "The Representative Regime" in the prestigious *Revue Bleue,* in the course of 1886 and 1887. His proposal was basically a presidential system, in which a small directorate would supervise government work, whereas the parliamentary assembly would limit itself to legislative tasks. Not surprisingly, he was to join the first seemingly strong leader who came along, General Boulanger.

Another radical politician, Gambetta's former assistant Joseph Reinach (1856–1921), however, expressed early fears that a presidential system would soon lead to a plebiscitary dictatorship. He agreed with President Grévy that the recent introduction of the "scrutin de liste" (which favored stronger groups and well-known personalities) was in reality the "scrutin des riches" and "the inaugural coach for princes and generals" (1888, in Reinach 1890, p. 188). He therefore became one of the main opponents of General Boulanger. The rise and fall of the movement around that particular figure, then, which domi-

31. *La Démocratie et la France* (pp. 23–5, 47–50). English translation: *Democracy and France* (pp. 21–4, 43 ff.).
32. See Prins, *La Démocratie et le Régime Parlementaire;* Droz, "La Démocratie et son Avenir" in *Etudes et Portraits Politiques,* among others.
33. Thomson, *Democracy in France since 1870,* p. 112.

nated the years 1886–9, may be seen as the perfect illustration of this whole debate. It also became one of the prime contemporary sources of inspiration for Le Bon's crowd psychology.

The first present-day scientist to have pointed out the possible relevance of the Boulanger episode was Robert Merton, in his introduction to the American reedition of the *Psychologie des Foules*. "It is all a leaf out of Le Bon's book," he said. He added, though, that it was only one among many examples quoted, and limited himself to a summary description of the episode (pp. xx–xxiv). If we take a closer look at the work, however, we see that Le Bon was careful to limit his references to just a handful of contemporary politicians anyway. And if we take a wider view of his various bestsellers of the 1890s, we see that there are quite a few direct or indirect references to this curious political figure.[34]

Since Merton, other historians of crowd psychology have only briefly mentioned the Boulanger episode. We will go over its five successive phases because the basic facts are not standard knowledge and are even left out of many history books. Recent studies (such as Sternhell's 1978 book) have demonstrated, however, that the Boulanger episode was a significant step in the collective learning process of caesarist and nationalist movements in France and abroad, which were often leftist in origin but later drew decisive support from the Right. I will sum up some of these lessons and show how they relate closely to Le Bon's crowd psychology.

During the first of the five phases of Boulanger's rise and fall, he was just an ambitious general with a suitable image. He was somewhat taller than average, reasonably handsome, blond-haired, and blue-eyed. He had graduated from the famous Saint Cyr military academy, had served in various colonies, was often decorated, and had had the good fortune of being wounded in the Franco-Prussian war, just before the crushing of the Paris Commune revolt. Thus, he had been directly involved neither in capitulation, nor in repression. Thereafter he continued to cultivate contacts with his former schoolmate Clémenceau and his radical republican friends. When the Opportunists' government ran into trouble in the mid-1880s and felt forced to make overtures to the radicals, therefore, they decided to push for Boulanger's appointment as minister of war. After a royalist demonstration, furthermore,

34. See f.i. *Lois Psychologiques*, pp. 160–1; *Psychologie des Foules*, pp. 17, 40, 64, 109; *Psychologie du Socialisme*, pp. 48, 50, 469 – to mention some of the most obvious allusions.

the new government expulsed the various pretenders to the throne, and purged the army command.

Thus began the second phase: that of the popular minister. Boulanger undertook a complete reorganization of the armed forces. He introduced numerous technical innovations: formed an improved engineer corps, introduced modern weaponry (the Lebel repeating rifle), improved communication methods (use of the telephone), and limited paperwork. And he introduced social innovations as well: reduced service time and canceled exemptions (for bourgeois gentlemen and priests), improved conditions, and humanized rules. Barracks were renovated, halls of fame introduced, sentry boxes painted in the national colors. His measures were widely publicized daily and nationwide by another new invention: a full-fledged professional press department. Boulanger toured the provinces, made appearances at patriotic meetings, and introduced the military parade on the national holiday. At Longchamp, on July 14, the cheers of "Long live President Grévy" were soon drowned out by "Long live General Boulanger" when he proudly appeared at the head of his troops, mounted high on a magnificent black horse.

Over the next few weeks, newspapers carrying his portrait saw their circulation boom, his brief vita sold a hundred thousand copies, songs and souvenirs began to appear. The conservative daily *Le Figaro* wrote that it all revived painful memories. He has understood, it said, that rather than an average man, France wanted to have a prince at its head, or at least a soldier. "And because the nation likes feathers, he has shaken his before the crowd." A few weeks later, it returned to the subject: "Instinctively, the crowd acclaims the man which it feels in possession of a strong will. . . ." One should not be surprised that "the crowd . . . transfers on him who personifies the army – today closely linked to the entire nation – the signs of confidence and affection that it displays for it."[35]

Yet, it was exactly this fusion of caesarism and nationalism that began to worry friends and enemies alike. Boulanger had meanwhile adopted the nickname "Général Revanche," and had made it abundantly clear that he would "liberate" the Alsace-Lorraine region from German occupation as soon as he could.[36] Chancellor Bismarck de-

35. July 25 and August 15, 1886, respectively. Quoted in Néré, *Le Boulangisme et la Presse,* pp. 28 and 40 (my translation).

36. Zeldin (1979–81, Vol. 3, pp. 77 ff.) shows how the true feelings of the majority of people in the region were grossly distorted in French accounts of those days.

cided to enlarge the army, and when the Reichstag refused to vote the corresponding law, he simply disbanded it, had new elections, and won them, too. In the spring of 1887, then, tension was rapidly rising. The French minister of war toured the border area and canceled all leaves, prepared for mobilization and even an ultimatum. President Grévy asked him: "Don't you know that means war?" Boulanger is reported to have answered: "But yes, I am ready."

Using the pretext of a quarrel over the budget, the "opportunist" republicans thereupon decided to topple the government, to get rid of some radical ministers again, and first in line was Boulanger. This maneuver had become necessary because nationalist groups such as the League of Patriots had meanwhile adopted Boulanger as their hero. Now, a third and intermediary phase began in which he was still an ambitious general, no longer a popular minister, but not yet a declared politician either – since military men in active duty were prohibited from seeking a parliamentary mandate.

As soon as Boulanger had been deposed, some radical journalists had started a campaign to add his name to that of their preferred candidate on the ballot in a by-election in Paris. Almost forty thousand people did, and the candidate triumphed. Over the next few weeks, the capital was buzzing with nationalist agitation, demonstrations, and meetings. At the approach of the fateful date of *Quatorze Juillet,* the government decided as a precaution to appoint the general to Clermont-Ferrand, a long distance from any major urban centers. A nationalist journal immediately informed its readers of this maneuver, however. A crowd of around fifteen hundred people gathered at the Hôtel du Louvre, where Boulanger was staying, then accompanied him to the Gare de Lyon train station; their number rose to eight thousand later that evening. Some people lay down on the rails to prevent his departure, but Boulanger decided to bide his time, and left only a few hours late.

Toward the end of that year, however, the tide seemed to turn. It was discovered that the "opportunist" deputy Wilson had accepted bribes to get his father-in-law, President Grévy, to grant certain people decorations. After some hesitation, the head of state was forced to resign. The radical leader Clémenceau blocked his succession by the hated "opportunist" strongman Jules Ferry – who was wounded a week later in an assassination attempt. Instead, he conspired with Boulanger and others to "vote for the stupidest": a rather mediocre politician from a respectable republican family, by the name of Carnot.

General Boulanger's send-off by a crowd of followers at the Gare de Lyon. (Top from J. Néré, *Le Boulangisme et la presse*. Paris: Colin, 1964. Bottom from J. de Lacretelle, *Face à l'événément: Le Figaro 1826–1926*. Paris: Hachette, n.d., p. 85.)

But even some left-wingers now began to have mixed feelings about Boulanger's ambitions. The socialist *Cri du Peuple* wrote, for instance: "In our popular quarters it is known by experience that force prevails over right! That is why, General, you have with you the population, this indecisive and floating mass which cries Long Live this or that. . . ."[37] Boulanger's admirers continued to urge voters to add his name to the ballot in several by-elections, and they often did. The government felt, however, that the general was encouraging such moves and thereby breaking the rules: During spring of 1888, he was first suspended, then dismissed. This made him a hero as well as a martyr, and freed his hands for a further political career.

The fourth phase started when an ad hoc Committee of National Protest quickly developed into a kind of directorate for a new mass movement. Its chairman was Boulanger himself; the vice president was the leftist senator Nacquet, an early critic of the parliamentary regime; and the treasurer was the businessman Dillon, who in turn maintained good relations with the Right. A groundswell of sympathy arose from a wide range of personalities, newspapers, and organizations, covering the whole political spectrum from Blanquist revolutionaries to monarchist restorationists – a truly unique feat in those days. Their platform was correspondingly vague: Dissolution (of parliament), Revision (of the constitution), Constituent (assembly for the founding of another regime). They opened a major propaganda offensive: with thousands of hired hands, which recruited tens of thousands of people for mass meetings, and distributed hundreds of thousands of Boulangist leaflets. People took to wearing red carnations to publicize their support, or brandishing bread rolls at the end of a stick (in reference to Boulanger meaning baker).

During the rest of 1888, the deposed general won one by-election after the other, often with sweeping victories. In January 1889 he was elected with almost a quarter of a million votes in the central Seine district. Tens of thousands cheered him outside the Paris restaurant where he was having dinner with his friends, some urging him to lead the crowd to the presidential Elysée palace ("It is five minutes to midnight . . ."). But Boulanger felt sure the political landslide would continue anyway and that there could be no doubt that *he* would be the national leader opening the upcoming Universal Exhibition, commemorating the centenary of the storming of the Bastille. In order to accel-

37. January 7, 1888. Quoted in Néré (1963), p. 99 (my translation).

erate the drift, he expressed his profound respect for property, the family, and the church, and criticized the anticlericalism of the republicans. By now, radicals as well as Opportunists had become thoroughly suspicious of his ambitions, and decided to stop him by all legal – and if necessary illegal – means.

The final phase opened when a new government was formed with a "tough" interior minister, Constans, who had long been involved in political trickery himself. He swore to get rid of this clown: E finita la commedia! He used the first available pretext to start legal proceedings against the League of Patriots, which provided the main army of Boulanger followers. At the same time, he began opening the General's mail, tapping his phone [!], and keeping him under constant surveillance. He had soon collected a thick file of incriminating material on the hero's love life, business deals, secret contacts, etcetera. Not enough to arrest him, but enough to scare him. Some of it was carefully leaked to the subject himself, who did just what his adversaries had hoped he would do: He panicked and fled abroad, on April 1. . . . His closest advisers immediately begged him to return ("To appear or to disappear, mon général, that is the dilemma"), but he refused. This permitted the government to proceed unhindered with the dismantling of the movement. The subsequent elections were carefully rigged so that the movement only got a mere forty seats.

In true soap-opera style, the unraveling began when the exiled general accused a lady supporter of not caring well enough for the thoroughbred horses he had been forced to leave behind. This Duchesse d'Uzès then began talking to a journalist, and revealed that during all these years of radical rethoric she and other monarchists had quietly invested millions into the movement – in exchange for a secret promise of a regime change – should Boulanger be elected.[38] A little later, Boulanger's mistress died of a serious illness, and the would-be caesar committed suicide on her grave. His former friend and later enemy Clémenceau proposed to inscribe on his tombstone: "Here rests the general who died as he lived – as a sublieutenant."

38. She revealed these facts to the Boulangist deputy Terrail, who in turn wrote a series of articles for *Le Figaro* under the pen name Mermeix, later published as a book, *Les Coulisses du Boulangisme* (1890). During the last stage of his career, Boulanger is also reported to have written a letter to the Czar for support: "Sire, have your eyes on me when the moment has come, for me, to act . . . I will destroy the plague of parliamentarianism . . . I will erect an unsurmoutable barrier before the pernicious doctrines of socialism" (The letter was published after the Russian Revolution in the Catholic daily *La Croix*, June 23, 1923, and quoted by Dansette; my translation).

Yet this sarcastic remark barely covered a sigh of relief. Parliamentary democracy had just narrowly survived, and the faith of republican politicians in the rational judgment of "the popular crowds" had been severely shaken. Clémenceau's radical colleague Reinach wrote about the "caesarist virus" persisted "because the crowd being a woman, it sings and applauds the trumpeter passing by."[39] The opportunist leader Ferry said that the working masses instinctively longed for a strong ruler. The Boulangist phenomenon was the result of "what contemporary scientists would call: an autosuggestion of the multitude," he said. The people of Paris could easily be fooled: They were "not so different from the credulous crowds which have . . . imposed the Lourdes frauds as authentic." They created their own myths. "In both cases, truth and probability matter little to these popular imaginations, transient and powerful, naively visionary, which at certain hours of trouble or expectance need to incarnate their passions and illusions in a living form . . . the parties have exploited a certain state of mind of the multitude."[40] Later, two Boulangist sympathizers wrote novels about the mass movement (and also got in touch with Le Bon): Paul Adam published *Le Mystère des Foules* (1895), and Maurice Barrès *L'Appèl au Soldat* (1900). By that time, the latter had become the first politician to point to the necessity of a "nationalist socialism" for France.[41]

This latter link becomes even more interesting, if we proceed from a mere description to a brief analysis of the Boulanger movement. It has often been said that Le Bon foresaw the national-socialist and fascist techniques of crowd manipulation of the twentieth century. But this reverses the chronological order. Sternhell has shown in his seminal study, *La Droite Révolutionnaire 1885–1914: Les Origines Françaises du Fascisme,* that the Boulanger episode (together with the Dreyfus affair and other subsequent developments), rather than being a mere *faits divers,* were decisive steps in the collective learning process of caesarist and nationalist movements, ultimately leading them from a Bonapartist to a protofascist formula of mass mobilization. In a way, it tested a new configuration of political elements, which – after this

39. Article of 1887 in Reinach, *La Politique Opportuniste,* p. 235. Note that Reinach was a later correspondent of Le Bon.
40. Speech of April 11, 1889, and article written in 1890 for the *North American Review,* but published later in the *Revue de Paris* of July 1, 1897. See Robinquet, ed., *Discours et Opinions de Jules Ferry,* Vol. VII, pp. 146, 171.
41. On the novels, see Nye (1975) and Barrows (1981). On Barrès's political ideas, see Curtis's *Three against the Republic* and Sternhell's *La Droite Révolutionnaire en France,* pp. 33–4.

initial failure – was to be improved and become extremely effective. In order to illustrate this, we will discuss (only) some similarities between the Hitler and Mussolini movements on the one hand and the Boulanger movement on the other.[42]

In the first place, fascism often arises in the aftermath of a lost war, or at any rate one with an extremely unsatisfactory outcome. So did the Boulanger movement. The Franco-Prussian war, and the forced cession of the Alsace-Lorraine region to Germany, had exacerbated national-chauvinist feelings in France, and "le général Revanche" thrived on them. Fascism also frequently arises during an acute economic crisis and a mounting threat to the established social order. This was the case in the France of that period, too. Ever since the war reparations to Germany in the seventies, the economy had stagnated. After the crash of the Union Générale bank in the early eighties, it went into a real depression, which was to last until the midnineties. During this period, there was a rapid rise in revolutionary agitation.

Such a situation of political, social, and economic crisis is often accompanied by declining confidence in the parliamentary system and in democratic solutions. In the France of the late eighties the Wilson/Grévy decorations scandal added further to a widespread distrust of republican politics. At the same time, accelerated cultural change has often undermined traditional values in large parts of the country and added to a widespread call for a return to law and order. In the case of Boulanger, it is remarkable that some of his largest victories were scored around Paris and in the newly industrializing regions of the North, for instance.

Under these circumstances, conservative groups also become increasingly aware that they may be unable to stem the tide by ordinary means, and tried to take out an insurance policy by secretly coopting seemingly progressive factions with a radical social rhetoric but primarily nationalist goals. In this case, the legitimist, Orleanist, and Bonapartist pretenders to the throne (and their Catholic allies) were fatally losing ground. They therefore tried to buy a share in the Boulanger movement. If we limit ourselves to those secret donations that are well-proven, the total amounted to almost six million francs in

42. A good analysis of the interrelation between several elements in the fascist configuration is to be found in Kühnl's *Formen bürgerlicher Herrschaft*, part 2. I hope to return to the subject of the "mass psychology of fascism" in the projected Reich chapter for the sequel of this volume.

two years. Corrected for inflation, this equals the cost of a modern presidential election campaign in France a full century later.[43]

This contradiction between radical rhetoric and conservative financing can be resolved only by a fundamentally antidemocratic and authoritarian internal structure – in which only the leadership is aware of the true goals of the movement and of the means employed. This fits in perfectly with the caesarist myth of the "providential" man. The inner circle systematically exploits the financial ascendancy of the movement in various ways. Hired hands are recruited from the large mass of the unemployed, to be assigned as propagandists, agitators, and heavies. During his various campaigns, Boulanger had thousands of such *camelots* working for him. Meanwhile, propaganda material is distributed on an unprecedented scale. It is estimated that in the course of one year, the Boulanger movement itself produced more than three million copies of printed matter itself, and spent more than a hundred thousand francs on stamps alone – incredible amounts for those days.

The momentum of such movements is further reinforced by the silent support and complicity of large sectors of the business community and the government apparatus. Boulanger, too, had the sympathy of key people inside the army and the police force. Various studies agree that on various occasions, he could have taken power if he had really tried. It is this context that permits a small activist faction to absorb rival groups and weld them into a powerful political alliance. The League of Patriots alone grew to have some fifty to a hundred thousand members according to the police, and some two hundred to three hundred thousand according to its own leaders: an unprecedented number for those years.[44]

It is only against this entire background, merely outlined here, that the prominent role of crowd psychology in these groups can be understood. According to Eugene Weber, in his book *The European Right,* the Boulangist League of Patriots "was the first of many movements organized not for electoral and parliamentary action, but for the mobilization and manipulation of crowds outside the established structure of parties and parliament, indeed, even against them."[45]

43. For the secret donations see: Dansette, *Le Boulangisme,* pp. 155–6, 172–9, 204, 232, 317–22, 387; and Pisani-Ferry's *Le Général Boulanger,* pp. 134–6, 231, etc. In 1981, this would have amounted to some forty million francs, more than Mitterand reportedly spent on his relatively sophisticated and successful presidential campaign that year.
44. The numbers are quoted by Sternhell (1978), p. 96.
45. In Rogger and Weber, eds., *The European Right,* p. 85.

Rather than being a clairvoyant prediction of fascist techniques, then, Le Bon's *Psychologie des Foules* reflected (among other things) on the first experiment with a protofascist movement, drew a number of lessons from it, and subsequently inspired Mussolini and Hitler (among many others; see the last section of this chapter). Instead of blaming the leaders for their deceit and authoritarianism, furthermore, he blamed the followers for their credulity and submissiveness. He also urged Boulanger's republican opponents to start studying crowd psychology, and use it. If they did not, the crowds would follow their own course, and the results would be disastrous.

We do not know for sure how Le Bon felt about the rise of Boulanger, but it is probable that he was skeptical from the start. On the one hand, he was just in the process of courting the favor of the republican elite, and was even having some modest success at it. On the other hand, he probably felt repelled by the plebeian rhetoric of the general, and by his initial popularity in certain leftist circles. There can be no doubt that egalitarianism in all its forms was Le Bon's main enemy: feminism, desegregation in the colonies, and most of all "the threatening invasion of socialism" at home, as he was to write in *The Crowd* (p. 69).

The role of this other major contemporary mass movement as a negative example has also been recognized by most studies on Le Bon. With the exception of Barrows, however, most historians of crowd psychology have treated it in vague and general terms.[46] The emergence of the socialist movement is all too often outlined as an entirely gradual process that had been going on (and was to continue) for decades. Yet, one should immediately add that there were a number of significant leaps, such as the one between 1889 and (in France) 1893–95. Also, the references often limit themselves to the French developments. Yet, the heart of the matter is that it was a simultaneous international breakthrough, which scared national elites all around. Finally, socialism was not just a political ideology, it was also one of the very first lasting mass movements in modern history, which demonstrated its strength through crowd events on an unprecedented scale. Let us take a quick glance at the successive phases of this breakthrough.

The first phase was the gradual fusion of various activist groups into a more or less coherent movement. I have mentioned the fact that France was at the time going through an economic depression, and had

46. Barrows' *Distorting Mirrors*, ch. 1 among others.

Police charging the first May Day demonstration in Paris, 1890. (From M. Perrot, *Les ouvriers en grève – France 1871–1890*. Paris: Mouton, 1974.)

witnessed a rapid rise in social agitation throughout the eighties. This agitation remained largely episodic and fragmented, however. In 1889, furthermore, it was temporarily drowned out by the euphoric fanfare of the Universal Exhibition. On this same occasion, there were not only hosts of international scientific congresses, but also meetings with foreign worker's delegations. One of these meetings took the fateful decision to found a Second International Workingmen's Association,[47] and to call on workers of all countries to strike and demonstrate for an eight-hour workday on the First of May 1890. These twin measures created the general framework within which socialism was to manifest itself from then on.

The second phase started when these propositions were put to a test. At first, both friends and enemies had remained skeptical.

But as the crucial date approached and preparations advanced, governments got quite nervous. In Paris the authorities tried to prevent the meeting by arresting leftist leaders, intimidating socialist agitators, and

47. The First International had been plagued by the ongoing quarrel between marxists and anarchists, and had been dissolved some time after the Commune defeat.

Table 3. *Estimated number of participants in the Labor Day demonstrations of 1890 in France*

City	Participants
Paris	100,000
Marseille	40,000
Lyons	40,000
Roubaix	35,000
Lille	20,000
Calais	15,000
Saint Quentin	15,000
Bordeaux	12,000
Reims	10,000
Angers	10,000

Source: M. Dommanget, *Histoire du Premier Mai* (Paris: Éd. de la Tête des Feuiles), pp. 132 ff.

reinforcing the troops in the capital. Nonetheless, a large demonstration did gather, marched to Parliament, and presented a petition. Other cities saw a large turnout, too. Dommanget's history of the demonstrations (1972, pp. 132 ff.) provides the numbers shown in Table 3.

Smaller crowds gathered in up to 125 other cities and towns. The movement continued for several days. Even more important was the fact that similar demonstrations took place throughout Europe (and indeed in some countries overseas). In London some 300,000 people met in Hyde Park (Engels observing: "If only Marx could have seen this. . . ."). In Germany, an estimated 10 percent of all workers went on strike in the major industrial centers. In the big cities of the Austro-Hungarian Empire and Italy, there was a large turnout as well. In the smaller countries, from Scandinavia to the Balkans, millions of proletarians also went on strike and demonstrated for better conditions.

It was probably the largest crowd manifestation that had *ever* taken place in world history. This unprecedented display of proletarian power and unity shocked the bourgeoisie. Simon, another grand old man of the Third Republic, wrote in the national daily *Le Temps:* "What is grave, is the fact of [their] having reached an understanding across frontiers, having adopted a text of common demands, a common way of proceeding; of having put in movement so large a number of persons belonging to the most diverse nationalities and professions. . . ." And a provincial daily warned that "consciously or unconsciously" the demonstrators might obey slogans coming from enemy capitals such as Berlin.[48]

When it was announced that the feat would be repeated in 1891, both secular and religious authorities felt forced to react. In Fourmies, French troops opened fire on a crowd, turning their new Lebel repeating rifles against human targets for the first time. They killed ten people, mostly children and youngsters. Here again, fear of the mob killed more people than the mob itself. In the Vatican, two weeks later, Pope Leo XIII issued his encyclical letter *Rerum Novarum,* which criticized excessive exploitation and inhuman conditions but defended private property and social inequality. Christian trade unions were set up throughout Europe to try to co-opt the workers' movement and steer it from its revolutionary course.

In 1892, when May Day fell on a Sunday, it attracted strikers and nonstrikers alike. In most countries, the crowds were larger than ever: A half million demonstrators were reported in London alone. In France, however, the emphasis shifted back to politics. This brings us to the third phase of the socialist breakthrough there.

During the municipal elections of the same year, socialists scored major victories in several larger cities and towns. During the parliamentary elections of 1893, socialists of all persuasions more than tripled their share of the vote and won more than thirty seats (fifty if we include the socialist radicals). Everyone was well aware that this probably announced a larger shift in the long run. Within a half dozen years the first socialist did indeed enter the cabinet, and within more than a dozen years their alliance with the radicals reached a parliamentary majority. (The evolution of the parliamentary Left over these years is shown in Table 4. Note that three hundred seats were a majority.) The premonitions and fears of Le Bon and other conservatives, therefore, were not farfetched. But let us turn from this short description of the socialist breakthrough during this period to some analytical remarks on its wider significance.

1. The reason why this mass movement seemed so threatening to the national elites was that it was at once antinational and antielitist. It thrived on the hopes of the proletariat, which already included millions of people in France, tens of millions throughout Europe, and which

48. Dommanget, *Histoire du Premier Mai,* pp. 142 and 132, respectively.

Table 4. *Evolution of the parliamentary Left in France around the turn of the century*

Seats	1893	1898	1902	1906
Socialists	33	57	43	74
Radical socialists	16	74	104	132
Radicals	122	104	129	115
Total	171	235	276	321

Source: D. and M. Frémy, *Quid 1978* (Paris: RTL/Laffont, 1977), p. 578.

was bound to grow further both in absolute and in relative terms. Insofar as it was linked to any one country in particular, furthermore, it was linked with Germany – France's archenemy.

2. The movement possessed an elaborate ideology embracing economic, social, political, and philosophical theories. It is true that the "German" theories of Marx and Engels only circulated in one of the French factions before 1893, but they quickly spread to others thereafter.

3. The movement was gradually developing an organizational infrastructure, on a national as well as on an international scale. Initially there were at least four socialist factions, around leaders (*meneurs*) such as Vaillant, Guesde, Allemane, and Brousse. The Guesdist party, which was the strongest, reached ten thousand members in 1893 (a fivefold increase in four years). After their common breakthrough of that same year, however, three of these factions formed one parliamentary group with "independent" socialists such as Jaurès – who gradually succeeded in uniting the parties and emerged as their common leader. The electorate of the socialists (including the socialist radicals) reached six hundred thousand at that time, but their German counterpart already scored three times that number.[49]

4. The entire mass movement was gradually equipping itself with a highly elaborate and suggestive symbolism, underlining its militantly antielitist and antinationalist character. The First of May Labor Day soon became an annual festival, marking the end of the defensive

49. Guesdist party: Mayeur (1973), p. 181. French vote: Barrows (1981), p. 17 (= 5%). German vote: Kuczynski Vol. 4, (1967), p. 165 (= 23%). Just before the First World War broke out, the unified French Socialist party reached a level of a 101 deputies, 90,000 members, and 1.4 million voters (Abendroth 1972, p. 65).

Cover and score of "L'Internationale." The solidarity anthem was first performed in 1888 by workers in Lille. (From M. Perrot, *Les ouvriers en grève – France 1871–1890*. Paris: Mouton, 1974.)

winter period, and the beginning of the spring offensive for better conditions. For the working class, it soon outstripped the national *Quatorze Juillet* holiday in importance. The Red Flag (sometimes combined with black) came to symbolize the readiness to spill one's blood (or if necessary, face death) in the course of the struggle. For the working class, it soon superseded the national *Tricolore* flag as a symbol of group loyalty. In the course of the nineties, "L'Internationale" emerged as a hymn of proletarian solidarity across national frontiers, and for them came to equal in significance the national anthem, the Marseillaise. The emergence of these various symbols (described in successive monographs by the historian Dommanget, among others), coupled with the lofty aspirations for an ideal society, gave the early socialist movement every semblance of a religious faith, as many insiders and outsiders were quick to point out.

5. The preferred means of expression of this emerging mass movement were crowd events such as demonstrations and strikes, which underlined the fact that individual weakness might easily be turned into collective strength. The fourteen Bourses de Travail in the larger cities formed a federation in 1892. The next year, there were already some two thousand unions in industry and trade, which grouped a total of four hundred thousand people (a tripling in three years). That same year, there were already more than six hundred strikes, causing the loss of an unprecedented three million workdays (a doubling in three years time). By the mid-1890s a number of trade unions had formed the general confederation of labor (CGT), with its leaders Pelloutier and Briand even propagating the idea of the general strike as a revolutionary weapon.[50]

For all these reasons, then, Rébérioux speaks of an "explosion" of socialist activities in France between 1889 and 1895 (in Droz 1974, p. 160). Many bourgeois feared a national demise and a socialist takeover within the not too distant future. Yet another series of events were to exacerbate these fears.

The third widespread movement immediately preceding the publication of Le Bon's bestsellers was the sudden eruption of antiparliamentarianism between 1892 and 1894. This was stimulated by the revelation of widespread corruption among republican deputies, and expressed most dramatically in an anarchist campaign of violent attacks on selected political leaders and randomly chosen "bourgeois," the combination of which seemed to push the fledgling representative democracy to the brink of panic.[51]

The corruption originated in the wheeling and dealing that followed the stabilization of the republic in the eighties, the emergence of a national market and the related undertaking of large-scale infrastructural projects both at home and overseas. The Suez Canal, inaugurated in the late sixties, had dramatically changed the balance of power in the world, by opening up the East for further colonization. Some were intent on repeating this feat. Designer De Lesseps, born in 1806 and well over seventy, seemed vital enough to recommence – since he had

50. Kuczynski (1967), Vol. 33, pp. 130–2.
51. Both elements had already manifested themselves during the 1880s, but now returned on an infinitely larger scale. Similar events also occurred in other countries but on a smaller scale. Think of the Banca di Roma scandal and the murder of the Italian king we mentioned in the chapter on Sighele.

recently remarried an eighteen-year-old creole girl and given her a large number of children. In the late seventies, then, a company was formed for the digging of a Panama Canal, attracting large numbers of French shareholders. By the late eighties, however, it gradually became clear that it constantly needed more money, and it persuaded the major republican newspapers and the republican parliamentary majority to change the law, and permit the emission of a large number of lottery bonds, attracting more small savings from the general public. It soon turned out, though, that this could not prevent the company from going bankrupt, leaving behind more than a billion francs in unpaid debts.

The investigation into the affair was slow and difficult. But by late 1892, Barrès's Boulangist newspaper *La Cocarde* and Drumont's anti-semitic newspaper *La Libre Parole* revealed that the company financiers had contributed millions of francs to the anti-Boulangist campaign and other campaigns of republican newspapers and deputies, in exchange for their political support. The courts summoned the now half-demented De Lesseps, the famous Eiffel, and others involved, condemned them, but then released them. One financier fled abroad, another (an uncle and father-in-law of the radical politician Reinach) died mysteriously. Several former prime ministers, more than a hundred deputies, and a host of respectable newspapers were mentioned as having accepted bribes, but only a handful were effectively prosecuted.

All this further hampered public confidence in the democratic parties: After the next election, only a minority of the representatives was reelected. Le Bon took the rise and fall of De Lesseps as further evidence of the fickleness of the crowd (1895/1963, pp. 71, 79, 80) and scorned the "imbecility of certain magistrates who did all to dishonour us before foreign countries, on a matter of a few thousand franc bills accepted by half a dozen needy deputies" (1898/1977, p. 312).[52] Their scruples seemed all the more misplaced, Le Bon felt, because the revolutionary assault on the democratic regime was well under way and becoming ever more violent. The main threat came from the sudden development of a campaign of anarchist attacks, which seemed to strike at the heart of political institutions one after the other, from the parliament to the presidency.

Already during the incidents of Labor Day 1891, several anarchists

52. My translation.

had been arrested and condemned; but a brochure defending them confidently announced that they were "yesterday, a tiny handful; today, an army, tomorrow, an innumerable crowd – going where Truth is, no more bothering about the fearful sniggering of the rich than about the gloomy indifference of the poor."[53] So the subsequent year, individual anarchists decided to stop talking and start doing things, to make "propaganda through deeds," to open a violent offensive against the Parisian bourgeoisie.

In the early spring of 1892, the anarchist Ravachol (Koenigstein by his father's name) started a series of bomb attacks to "sweep away [this] rotten society." He was caught, examined (by the criminologists Bertillon and Lacassagne), judged sane, condemned, and executed. In order to dampen further agitation, the government closed the Bourse de Travail, but that only worsened the situation. That same spring and summer, there were recurring riots in the student-populated Latin Quarter. Meanwhile, bomb threats became a fad: the Paris police archives contain some four thousand threatening letters received in the course of that year alone.[54]

In the spring of 1893 Ferry – grand old man of the "opportunist" republicans and now president of the Senate – died from the belated effects of an earlier attack on his life. In late 1893 there were major attacks on a company, an ambassador, and finally on parliament itself. The government, which had been drifting to the right anyway, seized the occasion to have the Chamber pass the so-called *lois scélérates,* limiting civil freedoms again. In early 1894, there was another indiscriminate attack on a bourgeois restaurant, followed by almost monthly incidents. Many newspapers began carrying a special section on "La Dynamite," and a major part of Bataille's criminological annual was devoted to such affairs that year. But the apotheosis of the terrorist campaign was still to come.

In mid-1894 the Italian anarchist Caserio stabbed to death the French president Carnot (a relative and sponsor of Le Bon). The event decided the criminologist Lombroso to publish a psychophysiological study of the anarchist delinquents caught so far. The political crisis was further aggravated when the president of the Chamber, who succeeded Carnot, felt obliged to step down again, after public questioning of his suitability for the job had continued for more than six months.

Thus, when Le Bon embarked on his *Psychologie des Foules* the

53. Maitron, *Ravachol et les Anarchistes,* p. 33. 54. Ibid., p. 13.

Assassination of President Carnot, a relative of G. Le Bon, by the Italian anarchist Caserio. (From Romi, *Histoire des faits divers*. Paris: Éd. du Pont Royal, 1962, p. 17.)

Portrait and handwriting sample of Caserio, analyzed by criminologists such as Lombroso. (From C. Lombroso, *Gli anarchici*. Turin: Bocca, 1894, pp. 74, 75.)

sense of an impending collapse of the regime was very widespread indeed. Le Bon felt the reasons for the crisis were twofold. First, there was the failure of the republicans to provide strong national leadership. And second, there was the onslaught of socialism – both in its "collectivist" and its "individualist" form. Socialism, he wrote somewhat later, would inevitably lead to "an absolutely despotic regime which would suppress all competition, give the same salary to the capable and the incapable, and incessantly destroy, by means of administrative measures, the social inequalities which arise from natural inequalities."[55]

Le Bon's psychology of peoples, crowds, socialism, and education

Since the *Psychologie des Foules* was by far the most influential of the early crowd psychologies, I have devoted quite a number of pages to an outline of Le Bon's early works and the intellectual and social background of his crowd theory. A few more observations still need to be made here, however.

The first is on the purpose of these works. At the outset of the 1890s, Le Bon apparently felt an urgent need to advertise himself as a social scientist and self-appointed adviser to the ruling class. Medical doctors played an increasingly prominent role in republican politics anyway, and gradually overtook the Catholic clergy as the most influential group in parliamentary circles.[56] Le Bon had long frequented the literary salons of Paris, but he now aspired to start a debating club or "think-tank" of his own. In 1892 he asked the psychologist Ribot to co-chair a monthly dinner meeting with him, the "Banquet des Vingt." The tradition was to last a great many years and to attract a rapidly changing but powerful group of people.

Its primary goal was obviously to bring together academic scientists and political figures, although others were invited as well. Gustave Le Bon was a relative and protégé of President Sadi Carnot, and Carnot's son was a frequent guest at the dinner meetings. Théodule Ribot, in turn, was a namesake of Alexandre Ribot, who became prime minister that year and was to return to the job no less than four times. Another frequent guest was Raymond Poincaré, at first only an ordinary minister, but later three times a prime minister as well, and in the end even a president. Among various other cabinet members was Gabriel Hanoteaux, a recurrent minister of foreign affairs. The list of invités was not limited to prominent republicans, however. It also included Prince Henri d'Orléans (the grandson of King Louis Philippe) and Prince

55. *Psychologie du Socialisme,* translated in Widener, ed. (1979), p. 126.

56. Zeldin (1979–81, Vol. 1, ch. 2, p. 23) quotes a figure of 72 medical doctors in the 1898 assembly, but cautions that the spreading of a corresponding "scientific spirit" should not be overestimated.

Roland Bonaparte (the nephew of the Emperor Napoleon). Later Le Bon further extended his influence through a weekly luncheon, to which we will return.[57]

The second remark concerns the place of the *Psychologie des Foules* in Le Bon's entire work. There can be no doubt that the book can be correctly understood only if it is seen as part of a longer series. It had been preceded by an equally small book on the psychology of peoples and was followed by somewhat larger books on the psychologies of socialism and of education. They all addressed the national crisis of those years and the persistent fears of another revolution. Nye has already pointed to the fact that Le Bon's previous works covered close to a thousand pages, including notes and references, whereas he now cut them down to a few hundred pages written in a clear, direct, and apodictic style.[58] These books were obviously meant to reach a larger middle class audience, to strengthen its conservative views – and they did.

Finally, there was a gradual shift in the scientific arguments Le Bon invoked: from predominantly physiological claims to predominantly psychological ones. His earlier works had already contained some psychodynamic notions, and his later works still contained neurophysiological ones, but the main emphasis shifted.

The book immediately preceding the one on the crowd, then, was *Lois Psychologiques de l'Evolution des Peuples* (1894b). It was dedicated to his friend Charles Richet, who had figured prominently in the hypnotic suggestion debates (although the book owed more to the psychological heredity and evolution/dissolution theories of Théodule Ribot, to whom Gustave Le Bon's next book was to be dedicated). Fragments of the book had appeared earlier in Richet's *Revue Scientifique*, but one fragment welcoming continuous struggle and recurrent wars as a perfect opportunity for character formation was immediately followed by a sharp rebuttal from the editor, and later omitted from the book version published by Alcan.[59] In the course of its successive reeditions, the book underwent various minor changes. One might add

57. *Les Déjeuners Hebdomadaires de Gustave Le Bon*, p. 4.
58. Nye (1975, p. 52) also points to his early attraction to aphorisms, in the writing of which he was to excel later.
59. See "La Lutte Guerrière des Peuples" and "La Lutte Économique des Races" in the *Revue Scientifique*. Apparently Le Bon intended to develop these fragments into his next book, but later backed away from it. Opposite the title page of the *Lois Psychologiques*, its sequel *Luttes Sociales et Guerrières des Peuples* was even announced as being "in print." Instead, the next book was to become the *Psychologie des Foules*.

here that most reeditions (and translations, for that matter) of Le Bon's books falsely claim to be identical to the original.

The *Lois Psychologiques* was meant to be a short synthesis of Le Bon's voluminous works of the eighties, intended to distill from them the laws governing the rise and fall of civilizations and, according to the introduction, to expose the negative role of "modern egalitarian ideas" on race, sex and class. He claimed races had relatively fixed characteristics of both a physical and a psychological nature. There were natural races and artificial ones, most civilized peoples belonging to the latter category. They were communities of interests, ideas, and sentiments. Inferior races had limited variability, superior races greater variability. Their elites generated a great many ideas, but only a few of these could be picked up by the masses.

These ideas resemble the "idées forces" of Fouillée, but "do not exert an influence until after a very slow evolution, they have been transformed into sentiments and have come in consequence to form part of the character. They are then unaffected by argument, and take a very long time to disappear. Each civilization is the outcome of a small number of universally accepted fundamental ideas. Religious ideas are among the most important of the guiding ideas of a civilization." Yet the European, and most of all the Latin, peoples, were threatened by the rapid spread of modern illusions. "The acquisition of a solidly constituted collective soul marks the apogee of the greatness of a people. The dissociation of this soul marks the hour of its decadence. The intervention of foreign elements constitutes one of the surest means of this dissociation being compassed."[60]

The people have become disoriented, and civilization is on the brink of collapse. Modern man "has lost his faith, and with it his hopes. The masses, grown excessively impressionable and changeable, are no longer kept in check by any barrier, seem fated to oscillate without intermission between the wildest anarchy and the most oppressive despotism." Under the Roman Empire, the last paragraph concluded, a similar situation had resulted in takeovers by bloodthirsty tyrants. "After them came the final catastrophe brought about by the barbarians. History always turns in the same circle."[61]

The book coincided with an increasingly gloomy mood among part of the national elite and had to be reprinted the very next year. Nye

60. *The Psychology of Peoples* (1894–99), pp. 235–6, 234.
61. Ibid., pp. 219–20, 229.

mentions a total of seventeen French editions and sixteen translations. This unprecedented success stimulated Le Bon to continue right away, with a sequel further elaborating one point: the psychology of crowds.

In the course of 1895, fragments of this sequel were once again published in the *Revue Scientifique,* the entire book was published later that year as *Psychologie des Foules* by Alcan, and translated the following year as *The Crowd.* Whereas the solid "soul of the race" was eroding, Le Bon said, the fickle "soul of the crowds" was gradually taking over.[62] The present era, therefore, could best be characterized as "The Era of Crowds" (1895/1966, p. 14). "Today the claims of the masses are becoming more and more sharply defined, and amount to nothing less than a determination to utterly destroy society as it now exists, with a view to making it hark back to that primitive communism which was the normal condition of all human groups before the dawn of civilization" (p. 16). "Civilizations as yet have only been created and directed by a small intellectual aristocracy, never by crowds. Crowds are only powerful for destruction. Their rule is always tantamount to a barbarian phase" (p. 18). Therefore Le Bon took it upon himself to teach the elites the psychology of crowds, so that they could try to steer them away from their fatal course.

I have already said that the *Psychologie des Foules* was dedicated to his friend Théodule Ribot, who had elaborated the evolution/dissolution theory in France (although the book owed more to the hypnotic suggestion debate in which Charles Richet had participated). I have also mentioned the fact that the book relied heavily on previous authors. Yet Le Bon only mentioned Tarde in passing, and feigned ignorance of the works of Fournial, Sighele and the other Italian predecessors.

Part I, "Minds of Crowds," was largely based on the application of hypnotic suggestion theory to collective behavior. Yet it emphatically avoided making any direct reference to contemporary theorists on the subject with whose works Le Bon was familiar: whether Richet, Charcot, Liébault, Bernheim, Regnard, or even Luys. It opened by saying that the crowd did not necessarily involve a large number of people or their simultaneous presence in one spot. It could be a smaller number

62. In my opinion the French term "L'Ame des Races" should be translated as "the Soul of the Race" rather than "the genius of the race" (1966, p. 3 among others). Note that the French "foules" can be translated both as "crowds" or as "masses." The latter translation is indeed used occasionally throughout the book.

in one spot (a committee), or a larger number dispersed over a vast territory (a nation). Psychologically speaking, a crowd was formed when "the sentiments and ideas of all the persons . . . take one and the same direction, and . . . a collective mind is formed, doubtless transitory, but presenting very clearly defined characteristics" (pp. 23–24). The causes of this transformation were threefold: the emergence of a sentiment of power, anonymity and irresponsibility that allows the individual to yield to his instincts; the contagion of sentiments and acts; and most of all suggestibility (p. 30).

> We see, then, that the disappearance of the conscious personality, the predominance of the unconscious personality, the turning by means of suggestion and contagion of feelings and ideas in an identical direction, the tendency immediately to transform the suggested ideas into acts; these we see, are the principal characteristics of the individual forming part of a crowd. He is no longer himself, but has become an automaton who has ceased to be guided by his will. Moreover, by the mere fact that he forms part of an organized crowd, a man descends several rungs in the ladder of civilization. Isolated, he may be a cultivated individual; in a crowd he is a barbarian. . . . (p. 32).

Note, by the way, that the concept of "the unconscious" has a rather different meaning here for Le Bon than it was to have for Freud, a fact that was often overlooked (even by the latter author himself).[63]

This whole causal mechanism, Le Bon said, led to a wide range of pathological phenomena in both the affective and the cognitive life of crowds. As far as their affective life was concerned, the symptoms were fivefold:

1. Impulsiveness, mobility, and irritability
2. Suggestibility and credulity
3. Exaggeration and ingenuousness of the sentiments
4. Intolerance, dictatorship, and conservatism
5. Morality: "A crowd may be guilty of murder, incendiarism, and every kind of crime, but it is also capable of very lofty acts of devotion, sacrifice, and disinterestedness, and acts much loftier

63. Le Bon meant primarily the "reflexes" in the sense of behaviorism (conditioned reflexes becoming unconditioned ones after having been repeated long and often enough). Psychoanalysis meant a much more dynamic concept of the "repressed" (certain impulses being rejected by consciousness). Even Freud himself, though, seemed to be only marginally aware of this fundamental difference (see the references to Le Bon in his *Group Psychology*).

indeed than those of which the isolated individual is capable" (p. 57).

As far as the cognitive life of crowds was concerned, Le Bon said, three symptoms stood out:

1. Ideas: "Whatever the ideas suggested to crowds they can only exercise effective influence on condition that they assume a very absolute, uncompromising, and simple shape. They present themselves in the guise of images, and are only accessible to the masses under this form" (pp. 61–2).
2. Reasoning power: "The arguments they employ and those which are capable of influencing them are, from a logical point of view, of such an inferior kind that it is only by way of analogy that they can be described as reasoning" (p. 65).
3. Imagination: "The images evoked in their mind by a personage, an event, an accident, are almost as lifelike as the reality. Crowds are to some extent in the position of the sleeper whose reason, suspended for the time being, allows the arousing in his mind of images of extreme intensity that would quickly be dispelled could they be submitted to the action of reflection" (p. 67).

In summary, then, the convictions of crowds assume a "religious shape": "This sentiment has very simple characteristics, such as the worship of a being supposed superior, fear of the power with which the being is credited, blind submission to its commands, inability to discuss its dogmas, the desire to spread them, and a tendency to consider as enemies all by whom they are not accepted" (p. 73).

In Part II of the book, Le Bon goes on to discuss the "Opinions and Beliefs [or Creeds] of Crowds." Its first chapter was a further application of the evolution/dissolution theory to nation building. Spencer was mentioned several times: the first time with a specific critical remark which, like so many others, was copied straight out of Sighele.[64] Le Bon once again avoided making any direct mention of previous authors who had elaborated the evolution/dissolution theory and with whose ideas he was familiar: Jackson, Ribot, Binet, Janet, and others.

The factors influencing the opinions and beliefs of crowds were twofold, he said. On the one hand there were remote factors, such as (1) race, (2) traditions, (3) time, (4) political and social institutions,

64. Compare Le Bon's book I, ch. 1, with Sighele's Introduction.

and (5) instruction and education. The former three elements were of seminal importance; the impact of the latter two was usually overestimated. On the other hand, there were immediate factors, such as (1) images, words, and formulas, (2) illusions, (3) experience, and (4) reason. "It is not by reason, but most often in spite of it, that are created those sentiments that are the mainsprings of all civilization – sentiments such as honour, self-sacrifice, religious faith, patriotism, and the love of glory" (p. 116). Therefore, all depended on the nature of the suggestion to which the crowds were exposed.

Crowds were always looking for leaders. These, in turn, "are especially recruited from the rank of the morbidly nervous, excitable, half-deranged persons who are bordering on madness. However absurd may be the idea they uphold or the goal they pursue, their convictions are so strong that all reasoning is lost upon them" (p. 118). Their means of persuasion are affirmation, repetition, and contagion; but these only work on the condition that the leaders are both prestigious and successful. In summary, then, there are two levels: "Above the substratum of fixed beliefs . . . is found an overlying growth of opinions, ideas, and thoughts that are incessantly springing up and dying out" (p. 147).

The third part of the book finally focused on "The Classification and Description of the Different Kinds of Crowds." On the one hand, there are heterogeneous crowds, Le Bon said, whether anonymous (in the street) or not anonymous (in an assembly). On the other hand, there were homogeneous crowds such as sects, casts, and classes. Curiously enough, he claimed that his book dealt only with heterogeneous crowds. But a closer look reveals that this is not true. He constantly altered the meaning of the word "crowd" to suit the examples, and many of the examples referred to sects and movements recruited from "homogeneous" castes and classes. His constant references to anarchism and socialism are cases in point. Although Taine, Sighele, Fournial, Tarde, and others had also done so, one of Le Bon's merits was precisely that he was much more consistent in stretching the word *Crowd* to cover mass movements.

Two chapters of this last part were devoted to "criminal crowds" and "criminal juries"; they added few new elements, just more examples. But the real focus of the work was on the last two chapters: those on "electoral crowds" and "parliamentary assemblies." In the chapter on "electoral crowds," Le Bon railed against the simplemindedness of the voting majority, the incongruence of many verbal programs, the power of the electoral committees, and the prestige of certain political leaders. Yet, he stopped short of a direct attack on "the dogma of universal suffrage" since "time alone can act upon it" (p. 183).

Similarly, in the chapter on parliamentary assemblies, he railed against the same vices: "Intellectual simplicity, irritability, suggestibility, the exaggeration of the sentiments and the preponderating influence of a few leaders" (p. 187). He said their legislative work presented "two serious dangers, one being inevitable financial waste, and the other the progressive restriction of the liberty of the individual."[65] Yet, here too, he stopped short of a direct attack on democratic representation, since it remained the ideal government for "those who form the cream of civilization," and the best means "to escape the yoke of personal tyrannies" (p. 200). To take this as a declaration of faith in democracy, however, as some do, is decidedly reading too much into it.[66] Le Bon was just enough of a realist to recognize that an open and frontal attack on the liberal system would make him an outcast and might even prove counterproductive.

In order to back up his claims, Le Bon sprinkled his book with historical anecdotes. But they were often distorted and quoted secondhand. Direct references were limited to a half dozen articles and a dozen books, most of them completely marginal to the central argument. Most examples belonged to either one of the three categories, furthermore:

1. The ancient history of the great civilizations of East and West. most of these examples were carryovers from his earlier works.
2. The origins of contemporary France (the Old Regime, the French Revolution, the Napoleonic Empire, and subsequent upheavals). Most of this came from Taine. Le Bon commended Taine as "the most remarkable of modern historians," but he could not keep himself from yet again belittling his debt by adding that the reason why Taine "has at times so imperfectly understood the events of the French Revolution is that it never occurred to him to study the genius [soul] of crowds" (p. 21).
3. Events of his own day. Among these the Boulanger episode, the Panama scandal, and the rise of the workers' movement stand out,

65. It is true, by the way, that the French assembly (with its some 600 deputies plus 300 senators) was decidedly very large, and made orderly deliberation occasionally difficult (also see note 78 of the chapter on Sighele).
66. See, e.g., Moscovici, *L'Age des Foules*, pp. 100 ff., 143, 205.

although he avoided references to concrete socialist and anarchist organizations and leaders.

The general conclusion of the whole book was that: "Judging by the lessons of the past, and by the symptoms that strike the attention on every side, several of our modern civilizations have reached that phase of old age that precedes decadence" (p. 204). The dissolution of the national soul was signaled by the sudden upsurge of crowd behavior. "The populace is sovereign, and the tide of barbarism mounts." And the last sentence again sounded an ominous warning: "such is the cycle of the life of a people" (p. 207). This second thin book, too, scored an immediate success. There were almost annual reprints until the Second World War.

It is important to see that Le Bon's seemingly abstract theories on crowd rule barely hid a very concrete political agenda: It was the *trait d'union* between his preceding book on the psychology of nations and his subsequent book, *Psychologie du Socialisme* (1898). This latter work was somewhat larger than the two previous ones and was intended as a complete discussion of the deeper causes for the rapid spread of socialism. The lower classes had always resented social inequality, he said, but modern developments exacerbated these feelings. The doctrine of socialism was of course completely nonsensical, but a scientific refutation would not help, since its appeal was basically rooted in pseudoreligious sentiments. "No apostle has ever doubted the future of his faith, and the socialists are persuaded of the approaching triumph of theirs. Such a victory implies of necessity the destruction of the present society, and its reconstruction on other bases. To the disciples of the new dogmas, nothing appears more simple. . . ."[67]

In the German and Anglo-Saxon world, Le Bon continued, the chances for a rapid victory of socialism were limited, because it did not square with either national ideas or national character. In the Latin world, however, it found fertile soil since statist and egalitarian ideas, as well as lack of initiative and willpower, were widespread anyway. Therefore, the number of misfits was steadily growing. Even more worrisome than the illusions of the masses were those of the elites: They might of course adopt some of the humanitarian rhetoric but should not allow themselves to be taken in by it. For the third time in succession, therefore, Le Bon ended with a grave warning: A weaken-

67. Translation in Widener, ed. (1979), p. 107.

ing of the ruling class had led to the downfall of Athens, Rome, and Byzantium – France and other contemporary nations risked following their example. Once again, the book hit a nerve, and went through a rapid succession of reprints. But there was more to come.

Around the turn of the century Alexandre Ribot, the past and future prime minister, headed a parliamentary inquiry into the urgent necessity of educational reform. Le Bon soon joined the debate with another polemical book, *Psychologie de l'Education* (1902). It claimed that education should not focus on memorizing facts but should rather foster vigorous attitudes. His earlier study of horse training had taught him how persistent drilling always produced the desired reflexes. Similar methods should be applied to young Frenchmen. "Education," the book's motto said, "is the art of making the conscious pass into the unconscious." National education should therefore focus on the strengthening of national character and the inculcation of national ideals: more in particular on a "Cult of the Fatherland" (p. 302). Where colleges and universities failed, the last chapter said, the military draft provided an excellent opportunity to correct weaknesses and to "pour steel" into the hearts of French citizens.

Nye has shown how Le Bon's ideas were soon picked up by the directors and teachers of the École de Guerre, rapidly spread through the army command after the Dreyfus affair, and contributed to major changes in military thinking on the eve of the First World War. Among other things, they inspired heavy emphasis on army morale and offensive tactics. But the human wave attacks during the First World War took a rather heavy toll, particularly among the lower class and colonial infantry.[68] After victory had been won, both the government and the Academy of Moral and Political Sciences nevertheless honored Le Bon for his important contributions to the final result. Long after that, his influence on French army commanders like Foch, Pétain, and De Gaulle remained notable. Various foreign military leaders adopted similar ideas.

Le Bon as a political psychologist

Throughout these years, Le Bon had not only been involved in psychology and physiology, but also in physics. From the mid-1890s on, he had joined the ongoing research on atomic energy, and had per-

68. Nye (1975, p. 146) refers to the employment of West African troops as cannon fodder by a close friend of Le Bon, General Mangin, nicknamed "the butcher of Verdun."

formed tests in his own laboratory (which were, it is said, to contribute to his later blindness). His books, *L'Évolution de la Matière* (1905; the evolution of matter) and *L'Évolution des Forces* (1907; the evolution of energy), were well up-to-date, and even led him to dispute Einstein's priority in later years.[69]

Meanwhile, he had become quite a wealthy and influential man. He had changed publishers by moving from Alcan to Flammarion, where he headed the "Library of Scientific Philosophy." It was to contain more than two hundred volumes, including some of the most important works of his day. It put him in close touch with many major intellectual figures. After the monthly dinner meeting that he had co-chaired with the psychologist Ribot, he started a weekly luncheon meeting, which he co-chaired with the physiologist Dastre. They, too, soon attracted a mixed crowd of well-known scientists, artists, and socialites. Since Le Bon held the intellectual capacity of ordinary women in low esteem, his female invitées mostly belonged to the highest levels of the European aristocracy: la princesse Marthe Bibesco, la princesse Marie Bonaparte, l'infante Eulalie d'Espagne, la duchesse de La Rochefoucauld, la duchesse de Marlborough, and their like.

Diplomats and ministers also attended in large numbers. If we limit ourselves to French presidents and prime ministers, these included Barthou, Deschanel, Flandin, Herriot, Tardieu, and most of all Aristide Briand, whom Nye describes as Le Bon's "closest friend among the *politiques*" (p. 86). His friend Raymond Poincaré, in turn, "embarked on a willful attempt to elaborate a nationalist myth to the French people" after he became prime minister, according to Nye (p. 110). Relations with Georges Clémenceau, cool at first, gradually warmed. Among the past, present, or future heads of state or government leaders from foreign countries who participated in these meetings at one time or another were Alessandri of Chile, Paderewski of Poland, Benes of Czechoslovakia, Bratianu and Ionesco of Rumania.[70]

Gradually, then, Le Bon reached the position he had always aspired to: that of a senior adviser to contemporary rulers. In 1910 he opened a second round of social-scientific bestsellers with his own equivalent of Machiavelli's *The Prince:* his *La Psychologie Politique.* "I asked my

69. Note that Le Bon had written earlier, that Jews had "never made even the faintest contribution to the edification of human knowledge" (see note 31 of chapter 5, on Tarde). Le Bon's political admirer Duverger reedited a brochure with articles relevant to Le Bon's claim: *Matière Égal Énergie C'est Gustave Le Bon!* Robert Nye's wife Mary Jo has written an academic study on the subject.

70. *Les Déjeuners Hebdomadaires de Gustave Le Bon,* pp. 4–7.

eminent friend, professor Ribot, to indicate to me the treatises of political psychology published recently," he said, but "his anwer taught me that none existed." "The only known veritable treatise on political psychology was published four centuries ago . . . [but] the larger part of the rules relating to the art of leading men as taught by Machiavelli have been unusable for a long time."[71]

In the age of mass democracy, Le Bon said, "an elementary wisdom tells us to adapt to what one cannot prevent. Elites should therefore adapt themselves to popular government, dam and channel the fantasies of the greater number, just as an engineer does with a torrent."[72] They should particularly dam and channel the trend toward socialism and desegregation. The only feasible way to achieve this would be to foster national chauvinism and racial purity.[73] "It is precisely on the defense of the notion of the fatherland – which implies an entire moral organization – that our efforts ought to concentrate," he concluded.[74]

The book was followed by others, *Les Opinions et les Croyances* (1911; opinions and beliefs) and *La Révolution Française et la Psychologie des Révolutions* (1912). Although they were all relatively successful, and always included some new ideas, they were basically an application of identical principles to adjacent fields. In order to enhance the impact of his precepts even further, Le Bon also turned to condensing them into incisive formulas, from *Aphorismes du Temps Présent* (1913) on. Yet, he as far from finished.

Despite his old age, he published ten more books: mostly a running commentary on current affairs, such as the First World War, the collapse of the Central and East European Empires, and the emergence of a new type of social movements – all of which, he felt, confirmed his early views.[75] He died in 1931. A bust marks his grave at the Père

71. (1910, 1926), pp. 9 and 4–5, respectively (my translation).
72. Ibid., p. 122 (my translation).
73. Already in his *Psychologie du Socialisme*, Le Bon had expressed the "intimate conviction" that in order to civilize Africa, the Europeans would have to reestablish slavery under another name (to avoid offending "the philanthropists"). The American Civil War had been a terrible waste, he said, since the abolition of slavery had led to a notable "mental regression" among the eight million blacks: "Their incurable laziness, their stupidity and their dangerous bestiality make them useless in a civilized society" (book V, ch. 4, sec. 1. In the edition of 1898/1977, pp. 367–68, my translation). In the original English edition (p. 378) the brutality of this entire fragment seems to have been toned down. Le Bon's ideas on the Jews are discussed in the next chapter of this book.
74. (1910, 1926), p. 370 (my translation).
75. *La Vie des Vérités* (1914); *Enseignements Psychologiques de la Guerre Européenne* (1916); *Premières Conséquences de la Guerre* (1917); *Hier et Demain* (1918); *Psychologie des*

Lachaise cemetery; streets were named after him in Paris and his native Nogent. But his influence extended far beyond his own lifetime.

Le Bon's biographer Nye writes that "If one bulks together the French editions of his total literary output of about 40 volumes, the entire number of volumes approaches half a million in French alone. . . . He was the supreme scientific vulgarizer of his generation."[76] Le Bon's admirers did not hesitate to compare him to Newton, Lamarck, Darwin, and Spencer. He did indeed play a key role in the fundamental paradigmatic shift of the 1890s, which Hughes has labeled "The Revolt against Positivism."[77] During that decade, a wide variety of thinkers, often with rationalist and determinist backgrounds themselves, rediscovered irrationalism and voluntarism. Le Bon's thought influenced a whole generation of younger thinkers who battered bourgeois complacency and "the illusions of progress." In the field of political theory, three names stand out: the Frenchman Sorel, the Italian Pareto, and the German Michels. Some of their key works were partly inspired by Le Bon, and particularly by his reflections on socialism and the ways to confront it.

Sorel started out as a Marxist socialist himself, but in later life became an advocate of monarchist restoration. Although mildly critical of some claims of *Psychologie du Socialisme,* his review in the *Revue Internationale de Sociologie* nevertheless called it "the most complete work published in France" on the subject, praise that Le Bon immediately added as a footnote to the preface of later editions.[78] He also returned the compliment by calling Sorel the most erudite of French socialists.[79] A number of years later, Sorel incorporated some of Le Bon's ideas on the primordial role of political myths in his famous *Réflections sur la Violence* (1908), and praised him as "the greatest French psychologist" – particularly because of the practical nature of his political advice.

We have already briefly referred to Pareto in the chapter on Italy, and mentioned that he started out as a progressive liberal but also turned

Temps Nouveaux (1920); *Le Déséquilibre du Monde* (1923); *Les Incertitudes de l'Heure Présente* (1924); *L'Évolution Actuelle du Monde* (1927); and *Bases Scientifiques d'une Philosophie de l'Histoire* (1931).

76. Nye (1975), p. 3.
77. In *Consciousness and Society.*
78. Sorel (1899), also in Le Bon (1898/1977), p. xiii.
79. Curtis, *Three Against the Republic,* p. 52.

increasingly conservative. He, too, wrote a critical review of Le Bon's book, but nevertheless immediately thereafter embarked on a closely related elaboration of his own theory of elite circulation, *Les Systèmes Socialistes* (Socialist Systems, 1902). Note that Le Bon, too, had stressed the inevitability of elite rule in all his major works, although he put more emphasis on mass incompetence.

We have also referred to Michels in the same chapter, and mentioned that he started out as a revolutionary socialist but became a fascist sympathizer in later life. The publication of his study on German socialism, *Zur Soziologie des Parteiwesens* (on the sociology of parties, 1911), which included his famous "iron law" of oligarchization, was immediately followed by a letter to Le Bon. In it he called himself Le Bon's student, adding that all he had done was simply apply Le Bon's crowd psychology to political parties.

Nye has extensively analyzed the close relation between the political thought of these four authors (1973, 1975, 1977). What seems to unite them is a consistent attempt to break through the modern opposition between Left and Right, socialism and nationalism, masses and elites, democracy and dictatorship. In doing so, they all helped lay the theoretical groundwork for the new ideologies and movements that emerged after the First World War.

This illustrates that Le Bon's influence went far beyond France, or even the Latin world, of his days. Major works were translated into (in alphabetical order) Czech, Danish, English, German, Italian, Lettish, Polish, Portuguese, Rumanian, Spanish, Swedish, and other languages. They were translated into Russian by a top general; into Turkish, Arabic, and Hindi by government ministers; and into Japanese by an admiring ambassador – who felt obliged, however, to challenge Le Bon's ideas on the inferiority of his race and to announce impending proof of its superiority.[80] Le Bon's ideas prodded the elites into action everywhere, albeit with varying results.

It is hard to come up with a compact definition of Le Bon's ideological position, although it is clearly wrong to consider him an ordinary democrat, as some authors do. Like Taine, he belonged to a social group that had originally favored a return to a constitutional monarchy of the Orleanist type, but reluctantly settled for a conservative republic. There can be little doubt, however, that he would have pre-

80. Motono (1914), pp. 15–20.

ferred indirect or restricted suffrage if that would still have been feasible, and that he basically opposed the democratic idea of "one man, one vote." Throughout his work, he freely advocated male chauvinist, racist, militarist, and reactionary ideas – consistently opposing almost every single social reform of his day. Yet his cynical views contain some sobering thoughts and even occasional penetrating insights on the potential flaws and weaknesses of parliamentary democracy, and the list of twentieth-century political figures who adopted at least some of his ideas covers the whole spectrum from the far Left to the far Right.

It is even claimed that Lenin (who was exiled in Paris for some time) later had a heavily marked copy of one of Le Bon's major books on his desk.[81] It is an established fact that Le Bon and the other crowd psychologists were often quoted in the crucial "revisionism" and "mass action" debates between the right and the left wing of the German Social-Democratic party of those days.[82] Nor can there be a doubt that Le Bon (partly through others) influenced many of the revolutionary syndicalists who originally advocated a general strike as a means of forcing change.

Mainstream Western leaders, too, adopted some of Le Bon's ideas. The American president, Theodore Roosevelt, for instance, met with Le Bon in 1914 Paris, and reportedly told him that his *Lois Psychologiques* never left him during his travels.[83] The French president Charles De Gaulle often referred to *Psychologie des Foules*. And President Giscard d'Estaing may have been well acqainted with the *Psychologie du Socialisme,* since his conservative father co-edited a brochure with extracts from it.[84] Most European nationalists were influenced by Le Bon in one way or another. In this respect it is curious that some of the ideas of Theodore Herzl, the founder of Zionism (who

81. Widener (1979, p. 41) refers to Le Bon's psychology of peoples in this respect but gives no source. Vlach (1981–82, p. 351) refers to his psychology of crowds, instead, and says the information was given in an interview with Bajanov – former secretary of the Soviet Politburo – which appeared in *Spectacle du Monde,* April 1978, p. 79. Since Lenin's library has recently been opened for scholarly research, I asked an Eastern European colleague to check. But he said there were no Le Bon books in it today.
82. For instance, by Kautsky in "Die Aktion der Masse." See Grunenberg, ed., *Die Massenstreikdebatte,* p. 253 ff.
83. Nye (1975, p. 88) suggests that this claim may have been somewhat enlarged by Le Bon himself.
84. See Blanchard and Giscard d'Estaing, *Gustave Le Bon et le Socialisme.*

also was in Paris at the time), correspond with those of Gustave Le Bon, a proponent of antisemitism.[85]

Yet, these politicians only adopted *some* of Le Bon's ideas on the psychology of peoples, crowds, socialism, and education. Fascist leaders, by contrast, often adopted his ideas *en bloc*. According to Nye, the Rumanian princess Bibesco, daughter of a major conservative leader, drew the attention of the future dictator Antonescu to the ideas of "Le Docteur Faust de la Rue Vignon" as early as the 1920s, and in turn drew the attention of Le Bon to this political figure who might "succeed where Boulanger failed" (1975, pp. 165–6).

The Italian *Duce* Mussolini said in an interview that he had read all his works and frequently reread his book on the crowd. He also wrote in his autobiography that he had been interested in crowd psychology since his school days, and "one of the books that interested me most was the *Psychologie des Foules* by Gustave Le Bon."[86] Mussolini was thus elated when Le Bon sent him complimentary letters and autographed copies of his books after he had taken power, and even asked him for a definition of democracy. "Democracy," Mussolini replied – probably trying to express what he felt Le Bon thought but never said in so many words – "is the [form of] government which gives or seeks to give the people *the illusion* of being sovereign."[87]

The German *Führer* Hitler was an even more dedicated pupil. Most present-day biographies agree that he was heavily influenced by Le Bon.[88] It has been verified that there was a copy of the German translation of the *Psychologie des Foules* in the Vienna Hofbibliothek that Hitler used to frequent at the time he was trying to work out his political ideas, and ardently read everything related to the psychology of peoples, crowds, and socialism. In an article subtitled "Attempt at the reconstruction of a plagiarism," Stein has demonstrated that large parts of Hitler's *Mein Kampf* are based on paraphrases of Le Bon's book, and has identified at least fifteen fragments that closely parallel

85. See *Der Judenstaat* (1896). P. Loewenberg's discusses some of the background in his chapter on Herzl in *Decoding the Past*, pp. 101 ff.
86. See Doise in Graumann and Moscovici, *Changing Conceptions of Crowd Mind and Behavior*, p. 79 and Gregor, *The Ideology of Fascism*, pp. 113, 160, 411, 417.
87. See Le Bon, *Bases Scientifiques d'une Philosophie de l'Histoire*, p. 291.
88. See, e.g., the three biographies by Fest, Gisevius, and Maser (Vol. I, p. 155, p. 38 and p. 183, respectively, in the Dutch translations).

each other.[89] Even today, some of Le Bon's books are frequently reprinted and distributed by extreme right-wing groups.[90]

Thus, although few of Le Bon's ideas were entirely original, and most of his work rests on a habile synthesis of other theoretical contributions, his influence extends far beyond the scientific fields of psychology, sociology, and political science. Le Bon cast his shadow over the political events of the entire first half of the twentieth century and even beyond. His ideas on the crowd remain commonplace today. Still, theoretical developments have come a long way since.

89. *Adolf Hitler und Gustave Le Bon.*

90. Pierre Duverger, the man behind the French circle "Les Amis de Gustave Le Bon" makes no secret of his ultrarightist sympathies. Alice Widener's American anthology of Le Bon's works was published by the ultraright Liberty Press, etc.

5

The era of the public: Tarde, social psychology, and interaction

It was in Paris in 1898–99 that Tarde published the latter two of his four articles on our subject, which were also to make up the bulk of his book, *L'Opinion et la Foule,* a few years later. It is true that Tarde had begun developing his central theoretical concept of imitation much earlier, and had even begun applying it specifically to crowds in two earlier articles, published in 1892–93. However, this chapter demonstrates that (1) he had not yet entirely completed his theoretical framework in the early nineties; (2) its application to crowds (and sects) in these two earlier articles was still somewhat hesitant and inconsistent; and (3) only its later application to publics (and opinion) was truly remarkable, since (together with other late works) it pointed to an entirely new approach to both familiar and novel phenomena.

If we look at Tarde's four "crowd" articles from this perspective, we may easily discern an evolution in this sense. In the earliest article, he remained caught in a vague interpretation of imitation as suggestion, and of the crowd as a mostly physically assembled group. The second article widened the scope somewhat. But only in the last two articles did he really hit on a clear interpretation of imitation as interaction and the public as a physically dispersed group. Tarde was probably only partly aware of all the implications in this shift. But whereas he included both the first and the second article in a book published in 1895, he excluded the first and included the latter three in his book *L'Opinion et la Foule,* published in 1901.

A few selected quotes may suffice to highlight his gradual departure from previous theorists. "The expression *collective psychology or so-*

cial psychology," Tarde said in the opening lines of his preface to the latter book,

> is often understood in a chimerical sense, which must be discarded before anything else. It consists of conceiving a *collective spirit*, a *social consciousness*, a *we*, which is supposed to exist besides and above individual spirits. In our point of view, we have no need whatsoever of this mysterious concept in order to make a very clear distinction between ordinary psychology and social psychology – which we would rather call *inter-spiritual*. Whereas the first, in fact, deals with the relations between the mind and the totality of other beings outside, the second studies – or should study – the mutual relations between minds, their unilateral and reciprocal influences – first unilateral, then reciprocal. (Emphasis in the original)[1]

This quote clearly reflects his latter-day reading of imitation as interaction.

In the opening lines of the first chapter, Tarde added:

> Not only does a crowd attract and exert an irresistible pull on the spectator, but its very name has a prestigious attraction for the contemporary reader, encouraging certain writers to use this ambiguous word to designate all sorts of human groupings. It is important to put an end to this confusion, and notably not to confuse the crowd with the *public*, a word in itself subject to various interpretations but which I shall attempt to define precisely. We speak of the public at a theater, the public at some assembly, and here the public means crowd. But this is neither the sole nor even the primary meaning, and while the importance of this type of public has declined or remains static, the invention of printing has caused a very different type of public to appear, one which never ceases to grow and whose indefinite extension is one of the most clearly marked traits of our period.
>
> There is a psychology of crowds; there remains to be developed a psychology of the public, understood in this sense as a purely spiritual collectivity, a dispersion of individuals who are physically separated and whose cohesion is entirely mental. Where the public comes from, how it arises and develops; its varieties and relationships with those who are its directors; its

1. *L'Opinion et la Foule*, Avant Propos, p. 5 (my translation).

> relationships to the crowd, to corporations, to states; its strength for good and evil, and its ways of acting and feeling – this is what we plan to investigate in this study.

And he concluded: "I therefore cannot agree with that vigourous writer, Dr. Le Bon, that our era is the 'era of crowds.' It is the era of the public or of publics, and that is an entirely different thing."[2]

What we will do in this chapter, then, is explore the background of Tarde's special position among late nineteenth-century psychologists and sociologists in France, as well as the events between 1894 and 1901 that encouraged him to develop his ideas further. His transfer from a distant province to the bustling capital in 1894 stimulated a discussion with Durkheim over the relationship between psychology and sociology, and forced him to elaborate and refine his ideas on a possible social psychology. And, through his confrontation with the popular press, the widespread xenophobia of the time, and the Dreyfus scandal, he discovered the power of modern public opinion. That phenomenon had of course long been identified (see for instance Noelle-Neumann 1984), but had not yet been closely observed and analyzed in its most massive modern form.

Tarde as a provincial judge and criminologist

Gabriel Tarde descended from a prominent family of military officers and civil servants in the Périgord (Dordogne). He was born in the small town of Sarlat in 1843 and spent most of his life there. The old family manor still stands in the nearby village of La Roque-Gageac.[3] The family name had long been *de Tarde*, but the aristocratic prefix *de* had been abolished by the French Revolution. It was reinstated almost a century later, but Gabriel Tarde did not use it (though his sons did). It then seems incorrect to call Tarde aristocratic, as some authors do. He may best be considered a patrician: somewhat elitist, but far less militantly so than Le Bon and many other contemporaries.

Tarde's chief biographer Milet adds that the region had a long humanist tradition, and that this "vieux fond gaulois" had not been affected by the great upheavals of those days in the same way as Paris,

2. Translated in Clark, *Gabriel Tarde on Communication and Social Influence*, pp. 277 and 281, respectively.
3. One early occupant had been the astronomer/canon Jean Tarde (1561–1636), a friend and colleague of Galileo.

the north, and the east were. The downfall of the Second Empire, the lost Franco-Prussian war, the abortive Commune revolt, and even the Boulanger and Panama episodes always remained somewhat distant events here. This may help explain both Tarde's "eccentric" position in late nineteenth-century French social philosophy and social science, and the impact of the Dreyfus affair on his later thought – it was the first major political drama he witnessed from nearby.

Tarde was a Catholic, but in a very personal way it seems. He was not involved in the contemporary debates over clericalism. As far as his character was concerned, he had both introvert and extrovert traits. His father had died when he was only seven. In the course of his youth, he developed a very close relationship to his mother, and resented the Jesuit boarding school to which he was sent (calling it a "prison of innocents").[4] He was a brilliant student but developed several recurring illnesses, including asthma and an eye disease. At times, these kept him from reading and forced him to spend long periods ruminating previously absorbed ideas. These characteristics long dominated his personality and thought. After he had worked out his own approach and received some recognition, however, he eagerly joined both the literary salons and the debates of his day.[5]

Tarde studied law in Toulouse and Paris, then returned to Sarlat as a deputy judge's assistant. He gradually rose through the ranks to become a full judge there in 1875, as his father had been. He married soon thereafter and was to have three sons. So far, nothing destined him to become an original thinker. In private, he had already elaborated some far-flung theories. But he clearly lacked a stimulating intellectual environment in which to test and develop them.

Some time in the late seventies he wrote an anonymous (!) letter to Théodule Ribot, proposing a philosophers' association that would make a regular sojourn to the countryside and form "an academy in the ancient sense."[6] Ribot published parts of the letter, inviting the author to identify himself, and to submit his ideas to the *Revue Philosophique*. From 1880 onward, Tarde became a regular collaborator of the review. By the mideighties, his criminological contributions in turn attracted the attention of Alexandre Lacassagne, leader of the Lyons school,

4. Milet, *Gabriel Tarde et la Philosophie de l'Histoire*, p. 62, n. 4.
5. Certain details of Tarde's life and work were taken from obituaries by Espinas, Lacassagne, Picavet, A. de Tarde, and others.
6. Picavet, *Notice Biographique sur J. G. de Tarde*, p. 35.

who had just founded the *Archives d'Anthropologie Criminelle*. Thereupon, Tarde became a regular collaborator of the Archives as well, and later even a co-director. It turned out that Tarde was more of an abstract thinker than Lacassagne himself (who was also five years younger), and soon became *the* major spokesman for the French in their escalating dispute with the Italians.

During these years there were also major debates on crime and punishment in France. Nye has shown in an excellent study (1984) that there was a slow shift away from biological determinism, and toward psychological explanations. Tarde had been in touch with Lombroso, Ferri, and others since the early eighties, but soon proved to be quite critical of their theories. He reminded them that physical anthropology and social statistics had been inventions of "the Frenchmen" Broca and Quételet; compared Lombroso to coffee, "which stimulates but does not nourish"; called his *L'Uomo Delinquente* the "sterile work of an entire life"; and said *Il Delitto Politico* was fragmented, chaotic, and contradictory.[7] At one point, he even attempted a statistical refutation of the thesis of a "born criminal." While strongly rejecting evolutionary atavism as an explanation, Tarde became quite infatuated with hypnotic suggestion for some time.

We have seen in the previous chapter that the late eighties and early nineties displayed a rapid rise in interest in this phenomenon. Lawyers soon applied it to the problems of jury trials and criminal complicity alike. Tarde and many of his colleagues wrote various articles and papers on crimes in which suspects claimed they had been induced by hypnotic suggestion.[8] McGuire (1986d) has pointed out that the Charcot position was defended in the criminological world by the Swiss Delboeuf, whereas the Bernheim position was defended by the Belgian Liégeois. If the subject had still largely been absent at the Rome congress of 1885, it came to dominate the Paris congress of 1889 and those that followed. Crowds were, of course, held to be particularly susceptible to criminal suggestion.

Tarde's early articles on criminology were collected in two successive books, *La Criminalité Comparée* (1886) and *La Philosophie Pénale* (1890). The latter[9] (and another work published that same year)

7. In the *Archives d'Anthropologie Criminelle*, Tome II, p. 33 (quoted by Milet 1970, p. 23); *Actes du Troisième Congrès International d'Antropologie Criminelle*, p. 335 (quoted by Nye 1975, p. 114); *Revue Philosophique*, 19:611 (quoted by Milet 1970, p. 22); and the *Revue Philosophique*, 30:337 ff., respectively.
8. See also Tarde (1889), "Le Magnétisme Animal."
9. *La Philosophie Pénale*, pp. 320–3.

already contained several pages on crowd phenomena. We have seen in an earlier chapter how Tarde inspired Sighele, and how Sighele in turn inspired Fournial (a pupil of Tarde's friend and colleague Lacassagne), and finally Tarde himself. We therefore need not return to the circumstances under which Tarde's first article on the crowd was conceived. The Third International Congress of Criminal Anthropology, held in Brussels in the summer of 1892, featured few papers on "the born criminal," but quite a few on criminal suggestion. Tarde's paper, "Les Crimes des Foules," fit in perfectly with this trend.

"How does a crowd form?" he asked. "By virtue of what miracle do so many people – once dispersed and indifferent to one another – develop solidarity, aggregate into a magnetic chain, shout the same cries, run together, act concertedly? By the virtue of sympathy, the source of imitation, and the vital principle of social bodies."[10] Although Tarde's system focused on imitation, this first paper and article came close to identifying it entirely with suggestion. In line with earlier works by Sighele and Fournial, and the later work of Le Bon, he insisted on a regression supposedly occurring in crowds. Like them, he devoted ample attention to the supposed inhibition of higher, and reinforcement of lower, mental functions. Like them, he also devoted ample attention to the supposed differentiation and mediation of this process by physical, biological, psychological, and social factors. By contrast, he devoted relatively little space to mutual influence, and placed strong emphasis on the one-sided nature of criminal suggestion. There are always those who lead (*meneurs*) and those who are led (*menés*), he said, and penal responsibility should be assigned accordingly. In this sense, Tarde's oft-quoted first paper and article on the criminal crowd was rather atypical, and demonstrated little of his unique approach.

Yet, there were already isolated hints of his later turn of mind, as, for instance, in his discussion on the relation between crowds and "sects." "A handful of leaders awakens this latent power and directs it to a certain rallying point," he continued, "but – for this initial impulse to be followed and for this embryo of the crowd to grow quickly – a previous work, basically very similar, should have been accomplished in their minds. A slow contagion from spirit to spirit, a silent and tranquil imitation, has always preceded and prepared these quick contagions, these noisy and swaying imitations that characterize popu-

10. "Les Crimes des Foules," In *Actes du Troisième Congrès International d'Anthropologie Criminelle*, p. 73 (my translation).

lar movements." The work of crowds, he concluded, is always prepared by the work of sects.[11]

Tarde further developed this idea in his second article on the subject, "Foules et sectes du point de vue criminel" (crowds and sects from a criminal viewpoint), published the following year. Riots were often a by-product of crowd events organized by the mainstream socialists, such as strikes, demonstrations, or meetings. But other acts of political violence did not involve physical assemblies of people at all. Marginal activists such as the Blanquists and the anarchists, for instance, saw themselves rather as an avant garde. They felt small groups or even individuals could act as a catalyst in bringing about revolution. In these cases, collective persuasion was of a different nature, Tarde felt. "On the one hand," he affirmed, "a crowd tends to reproduce itself on the first occasion, to reproduce itself at more and more regular intervals, and – purifying itself each time – to organize corporatively in a kind of sect or party; a club starts out open and publicly, but little by little it closes and contracts; on the other hand, the leaders of the crowd are most often not isolated individuals, but *sectarians.* Sects are the yeast of crowds. Whatever serious or grave thing a crowd accomplishes, good or bad, has been inspired by a corporation" [association].[12] In this case, Tarde said, one may indeed claim metaphorically that certain associations are "born criminals," since they have been formed with the purpose of committing crimes. Others, however, are only "occasionally criminal," since they start out with a noble goal, which is then perverted. In both cases, these sects are often subject to the same mechanisms of regression that characterize crowds.

Yet, in the article Tarde was already placing less emphasis on suggestion and more on imitation proper. He also pointed occasionally to even wider phenomena: "a preparation of souls by conversations or readings" (p. 378), and the formation of opinion (p. 387). Tarde included both early articles on the crowd in a book titled *Essais et Mélanges Sociologiques,* published in 1895. The title indicated that he was trying to move beyond criminology, toward psychosocial science in general. He had meanwhile published a first summary outline of an entirely new approach that was well received by specialists. It added to his modest fame and contributed to a major change – which was to resound throughout his later life and work.

• • •

11. Ibid., pp. 73–4 (my translation).
12. "Foules et Sectes du Point de Vue Criminel," *Revue des Deux Mondes,* p. 370 (my translation).

For some time, Tarde's friends and admirers had been trying to find him a more prestigious job. In late 1893 the minister of justice asked him to prepare a memorandum on the reorganization of crime statistics in France, and in early 1894 even invited him to come and head the statistical department of the Ministry of Justice. Since his beloved mother had died a few years before, Tarde felt free to accept the offer. At fifty-one he left his native Sarlat and moved to Paris.

In this regard it is important to point out, that Tarde was not just a philosopher: He was also a mathematician. Statistics are a "sociomètre," he said in an oft-forgotten phrase.[13] Like his French contemporary Durkheim, he felt they were the key to psychosocial science. However, they differed strongly over the specific uses of statistics, and over their applications in sociology.

Tarde's years in Paris, then, were dominated by two developments. Intellectually, he embroiled himself in an escalating debate with Durkheim over the nature of sociology and its relation to psychology. This forced him to further elaborate his own theories and to apply them to proper contemporary phenomena, such as the emergence of modern public opinion. Politically, Tarde was confronted with a dramatic development in that domain: the Dreyfus affair, which was to dominate French public life for several years.

Sociologism, psychologism, and the Durkheim–Tarde debate

Tarde's arrival in the capital almost coincided with the beginning of a major debate on the new science of sociology in France (and indeed in the Western world as a whole). In this regard, most histories of sociology identify three major spokesmen and currents: Worms and the so-called biologistic approach, Durkheim and the so-called sociologistic approach, and finally Tarde and the so-called psychologistic approach.

René Worms (1869–1926) was to hold degrees in science, philosophy, literature, law, and economics. In 1893, when he was only twenty-four years old, he had begun to set up an elaborate infrastructure that was to allow social science to flourish. In the same year, he founded the *Revue Internationale de Sociologie,* the Institut International de Sociologie, and the Bibliothèque des Sciences Sociales. The very next year, the First International Congress of Sociology was held in Paris.

13. Lacassagne, "Gabriel Tarde," p. 526.

Worms succeeded in enlisting the support of many of the most famous social scientists both in France and abroad. In the following year, he revived the Société de Sociologie de Paris, which would hold monthly meetings.

Meanwhile Dick May (pseudonym for Mlle Jeanne Weill) had raised funds to found a Collège Libre des Sciences Sociales, an early predecessor of the present-day École des Hautes Études Sociales. Although this newly emerging little world grouped people from entirely different horizons, Worms himself long remained caught in the evolutionist paradigms of early sociology – witness his book *Organisme et Société* (1896). But he was soon faced with a formidable challenge by another early sociologist.

During these years, various psychologists and sociologists were trying to free themselves from their Darwinian/Spencerian roots. In French sociology, the most significant advances were made by Émile Durkheim (1858–1917). He was an assimilated Jew from the northeast, a background that proved highly significant during the Dreyfus affair, to which we will return. At the time he was lecturing at the University of Bordeaux. In line with the theoretical ideas of Comte and the statistical analyses of Quételet, he developed a functional approach to social phenomena. Since Napoleonic days, France had been trying to work out uniform administrative procedures, also yielding mutually comparable social statistics. The Third Republic reinforced this trend. By the early nineties long series of relatively reliable data on crime and death rates became available, contrasting various regions and causes, for instance.

In 1897 Durkheim published a pioneering study on suicide (in part based on data from Tarde's department).[14] It began with the observation (also made by Quételet and other predecessors) that regional and other differences remained fairly constant over time. Suicide, Durkheim concluded therefore, should not just be considered an abnormal phenomenon resulting from individual predispositions, but a normal phenomenon resulting from group processes (such as religious life). He distinguished egotist, altruist, and anomic suicide. Anomie, he said, is a social condition in which norms have disintegrated or disappeared, thus facilitating deviant behavior. This was the situation threatening (certain parts of) contemporary France.

14. The Italian Ferri also wrote a study on suicide and homicide. Note that suicide is often considered a sociological subject, whereas homicide is considered a criminological one.

Earlier, Durkheim had published another pioneering study on *Les Règles de la Méthode Sociologique* which argued sociologists' claims to a separate territory of exclusive expertise. Social facts, he maintained, must be seen as "things" exercising a collective constraint on individual beings. Social facts could only be explained by other social facts, and should not be reduced to psychological or biological facts.

Still earlier, Durkheim had discussed the social facts characteristic of primitive and modern societies. The former, he said, were characterized by a high degree of resemblance between individuals and a mechanical form of social solidarity. The latter, by contrast, were characterized by a high degree of differentiation, division of labor, and an organic form of solidarity. This modern society, however, still needed a common faith (in the French people, for instance), and a collective conscience to keep it from disintegrating. Durkheim, who held more or less liberal progressive political convictions himself, thus provided a tailor-made scientific ideology for the radical republican bloc, which was seeking to stabilize and consolidate the regime through nationalist education.

Durkheim quickly formed his own school around the journal *L'Année Sociologique* (1898), which soon came to surpass the eclectic Worms group, particularly for its influence in universities and lycées. In hindsight, however, Durkheim's "sociologistic" approach had various shortcomings. In his eagerness to proclaim the "independence" of sociology, he tended to sever *all* ties with psychology and deny the use of a mediating link in psychosociology or social psychology. His undue emphasis on collective consensus and social stability, furthermore, often blinded him to the ways in which new ideas and challenges to the existing order emerge.[15] Paradoxically, it was the seemingly more traditional approach of a somewhat more conservative thinker that avoided these pitfalls and pointed to a possible alternative. Its proponent was Tarde, the first new president of the Société Sociologique de Paris, and previously a key member of the Worms group.

Tarde had always aspired to being more than a small-town judge or even a well-known criminologist. In fact, he had already produced quite a number of literary and philosophical texts, but he published little before the 1890s, and some of what was published only appeared (in new versions) after he had moved to Paris. In the late seventies, for

15. See Coser (1971), Goddijn (1973), Szacki (1979), and others.

instance, he published a collection of poems and short stories, but soon changed his mind and bought back the remaining stock. In the eighties, he also wrote a few longer science fiction stories. One, *The Bald Giants,* was a reflection on the psychophysiological debates of the decade, depicting a society in which the head of each newborn infant was placed into a mold in order to produce the proper bumps (and therefore gifts) for science and play. The other, *A Fragment of Future History,* was a more direct reflection on sociological phenomena, describing a world in which the sun had become extinct and man was forced underground – with all the changes in social relations that implied. The two stories were not published until the nineties, however; the latter being widely acclaimed for its originality.[16]

Tarde was literally an eccentric in many respects. His thought showed more affinity with Anglo-Saxon individualism than with the various kinds of collectivism (socialism, nationalism) which came to dominate French republicanism (and even neoconservatism) in his day. He had more affinity with eighteenth- and early nineteenth-century French thinkers than with many of his contemporaries. In a way this cultural leeway helped him see through the shortcomings of certain contemporary ideas and lay the groundwork for overcoming them.

Like so many pioneers of his day, Tarde was well acquainted with the historical background of philosophy and recent developments in neighboring countries. His thought was clearly related to certain English, German, and French schools. His central notion of imitation, for instance, was basically an offshoot from a long tradition of British thinkers such as Smith, the Mills, and Spencer. They felt that individual freedom and the pursuit of happiness would automatically lead to social progress, because egotism would be held in check by "sympathy" and innovations would spread through "imitation." One immediate predecessor of Tarde in this respect was Bagehot, whose book *Physics and Politics* contained an entire chapter on the role of imitation in nation building. It was translated into French and underwent a rapid succession of reprints.[17] Tarde's tripartite mechanism of imitation, opposition, and invention, furthermore, was clearly reminiscent of the Hegelian scheme of thesis, antithesis, and synthesis.

16. The English translation was prefaced by H. G. Wells. It has been published again in recent years.
17. The original was published in 1872, the French translation in 1877. Sighele (1891, Intr.) referred to a fifth French printing of 1885. According to several authors, the Russian psychosociology of Mikhailovsky (1842–1904) contained similar elements.

Finally, his thoughts on the history of ideas were in turn inspired by a succession of French predecessors – all of whom shared a vivid interest in psychology, philosophy, logic, mathematics, and time series. The first of these was (de) Condillac (1715–80), author of *Essai sur l'Origine des Connaissances Humaines* (essay on the origins of human knowledge) and other works, who distinguished thoughts and passions, and already tried to formulate the laws governing their association. The second was Maine de Biran (1766–1824), author of *Essai sur les Fondements de la Psychologie* and other works, which emphasized the importance of ideas, and the primacy of inner life. The third and foremost influence was Antoine Cournot (1807–77), author of *Considérations sur la Marche des Idées et des Événements dans les Temps Modernes* (reflections on the course of ideas and events in modern times) and other works, which developed a probabilist view of social change. It was to the memory of Cournot that Tarde dedicated his first and most renowned study of psychology and sociology. Milet (1970) wrote a detailed monograph on the background of Tarde's "philosophy of history." A related contemporary French author, one might add, was Fouillée, who (also) emphasized the extraordinary role of *idées forces* in historical change. It should be noted, however, that Tarde took painstaking care in distinguishing his own thought from that of others.

As early as the 1870s, Tarde had embarked on devising a wide-ranging philosophical system of his own. He had written two manuscripts, *La Différence Universelle* and *Les Possibles*, and a book, *La Répétition et l'Évolution des Phénomènes*. Although they remained unpublished for some time,[18] they already contained the essential ingredients of his later thought. One may distinguish four axioms. First: The "real" emerges from the "possible" through contingency. Second: The "real" is articulated by resemblances and differences. Third: The philosopher and the scientist should study these resemblances and differences in their repetition. Fourth: The nature of the prevalent repetition is different in physics (vibration), biology (heredity), and sociology (imitation). Thus the social fact par excellence is imitation, and psychology and sociology should focus on discovering its laws.

In the course of the 1880s, Tarde attempted to do just that in various articles which he contributed to the *Revue Philosophique* and other journals. A major study by Lubek (1981) has shown that as early as the

18. Some of it returned in his work on *L'Opposition Universelle*, published in 1897.

mideighties, Tarde embarked on a pioneering work in two parts, to be titled *Psychologie Sociale et Logique Sociale*. The first part, however, was published in 1890 as *Les Lois de l'Imitation*. His first major noncriminological work, it rapidly established him as an original thinker. The second part was published in 1895 as *La Logique Sociale*,[19] and a popular summary followed in 1898, *Les Lois Sociales – Esquisse d'une Sociologie*.

Every now and then, Tarde said, new ideas occur: an "invention" is made. More often old ideas are copied: "imitation" is the rule. On occasion, however, certain people resist imitation, and "opposition" occurs. Thus society is characterized by the spreading of novelties (fashion) and the conservation of habits (custom). The social process is a continual interweaving of these various tendencies.

What Tarde meant may best be illustrated by the well-known water metaphors. On the one hand, one may think of smaller and larger drops falling into a pond, creating a pattern of widening circles, penetrating each other and disappearing again. On the other hand, one may think of tiny streams flowing into rivers, seeking a bed, eroding banks, pounding dams, etcetera. Society is like one large irrigation system: with currents, undercurrents, and countercurrents in constant flux.

The psychosocial realities to which these three forms of imitation apply are twofold, Tarde said. On the one hand, there is the cognitive aspect of beliefs (*croyances*). On the other hand, there is the affective/conative aspect of wishes (*désirs*). Both may be considered quantities: the latter force usually being stronger than the former. According to Tarde, these characteristics of the person correspond to various other elements. Clark (1969, p. 39) summarized this system as shown in Table 5. On the basis of this conceptualization, Tarde attempted to discover the laws governing imitation.

He said these existed on various levels. First, there were logical laws. The spreading of an innovation, for instance, may depend on a rational consideration of its practical uses. The elements are introduced in a "collective meditation," they are confronted in a "logical duel," and this may lead to "logical coupling" – an acceptance of the innovation (possibly in a somewhat adapted form). Second, there were extralogical laws. Imitation often conforms to psychological and sociological preconditions. Those who are closer to the source of

19. Milet (1970, p. 31) had *La Logique Sociale* published in 1893, Clark (1969, p. 6) in 1894. The title page of my copy bears the mention 1895.

Table 5. *Clark's summary of Tarde's system*

Probabilistic conditions	Personality characteristics	Social imitation patterns	Social beliefs	Basic cultural elements
Credibility	Belief	Credulity	Public opinion	Truth
Desirability	Desire	Docility	General will	Value

Source: T. N. Clark, *Gabriel Tarde on Communication and Social Influence* (Chicago: Univ. of Chicago Press 1969), p. 39.

innovation will adopt it sooner than those who are farther away. Those who possess prestige and power are more easily imitated than others, and so on.

In various articles and books, Tarde contrasted this approach with the "biologism" of Worms and the "sociologism" of Durkheim. Sociology, Tarde said, could only be built on (social) psychology. Durkheim, of course, did not fail to retort that this approach amounted to psychologism instead. For several years, then, their rapidly escalating debate marked the new field of sociology.

Both authors were trying to work out a foundation for this young discipline, and inevitably did so in opposition to contemporary views. Tarde was fifteen years older than Durkheim (and was to die thirteen years earlier). He was already middle-aged when he published his first work on (psycho) sociology in 1890. Durkheim, by contrast, was much younger when he published his doctoral thesis in 1893. It included a dozen – mostly critical – references to Tarde's work, and Tarde reacted with a critical review in the *Revue Philosophique*. Yet, the exchange was still courteous. In 1894 Durkheim added a substantial footnote on Tarde to the article that was to become the first chapter of his landmark book on sociological methodology. Tarde reacted with a rebuttal of Durkheim at the First International Congress of Sociology in Paris.

In October that year, they met for the first time. Afterward they exchanged letters, emphasizing that the debate was entirely scientific and not at all personal. In the spring of 1895, for instance, Durkheim wrote to Tarde: "I thank you for letting me know that this doctrinal disagreement does not diminish the esteem which you kindly choose to have for me; I do not need to tell you that, from my side, my deference for your work and person remains fully intact."[20] But the tone soon turned from irony to sarcasm. Tarde made highly critical references to Durkheim's notion of social facts in some of his articles and books of those days; Durkheim devoted an entire chapter of his famous book on suicide to a refutation of Tarde's notion of imitation. After 1898 (that is to say at the height of the Dreyfus affair), the polemic turned sour and divided the sociological world for years to come.[21]

For a long time, Tarde and Durkheim lived geographically rather close to each other, but apparently without getting together. Tarde remained in Sarlat (only a hundred miles east of Bordeaux) until 1894, when he came to Paris. Two years later he started lecturing at Boutmy's unofficial École Libre des Sciences Politiques (see Chapter 1, on Taine) and at Weill's similarly unofficial Collège Libre des Sciences Sociales. It was only in 1900 that he was made a professor at the official Collège de France (and a member of the equally official Académie des Sciences Morales et Politiques). His request to change the title of the chair from philosophy to psychological sociology was turned down, and upon his death four years later he was succeeded by a philosopher – Bergson.[22] Durkheim, on the other hand, lectured in Bordeaux until 1902, when he was made a professor at the Sorbonne in Paris. Within a year, mutual colleagues organized lectures and a public debate between Durkheim and Tarde – the culmination of their long-lasting dispute.

T. Clark noted that "the debate represented more than a simple disagreement of personalities; in it, two conflicting traditions of thought, two opposing sectors of French society, two hostile sets of institutions did battle" (p. 8). On the one hand, there was Cartesianism. "The legacy associated with Descartes was identified with reason, order, and authority, and housed in the bureaucratic institutions exemplifying this *esprit de géométrie*" – such as the governmental administration and the state educational system. On the other hand was the basic cultural configuration of spontaneity: "It was a mentality of artistic creation, romantic subjectivism, and personal invention guided by an *esprit de finesse*" (ibid.).

Tarde reproached Durkheim that his self-contained world of social

20. Milet in Tarde (1973), p. 24, n. 10 (my translation).
21. Van Heerikhuizen, Het Eigenaardige van de Sociologie, ch. 3a; Lukes, *Émile Durkheim*, ch. 16.
22. Milet (1970, pp. 26, 383 ff.) notes certain affinities between the thoughts of Tarde and Bergson.

facts and collective consciousness was a reification, and excluded social conflict. Durkheim retorted that Tarde's concept of imitation and his psychological reductionism were vague and contradictory. In a way, it was the age-old discussion over the relation between continuity and change, the whole and the parts. Each contended that the other twisted his message and used debater's tricks to get the upper hand. In fact, both overstated their cases and took up indefensible positions at one time or another.

There can be little doubt, however, that Durkheim came out the "winner." His influence on French sociology grew tremendously, whereas Tarde's impact diminished. This was partly due to the fact that Tarde had always remained a loner, lecturing only four years in a prestigious (but still somewhat marginal) official institution, whereas Durkheim's fifteen years at an equally official (but much more influential) institution enabled him to build a powerful network, which in the end penetrated the entire university system, the teacher's colleges, and the lycées. Lubek and others have summarized this contrast well by saying that Tarde may have developed an "exemplary paradigm," but unlike Durkheim, lacked the support of a "paradigmatic community."[23]

Lubek has also demonstrated how the outcome of this struggle has continued to affect social science even today. Between the mid-1960s, and the early 1980s, the *Social Science Citation Index* showed more than four thousand references to Durkheim, against less than two hundred to Tarde. Since the 1970s, however, there is a timid revival of interest in Tarde. This is largely due to the rediscovery of his role as a predecessor of social psychology, interactionist sociology, attitude studies, communication research, and most of all due to his pioneering views on opinion formation. The ideas of his later Paris years clearly show a further development in this direction, and this was closely related to the events of the day, particularly to the Affair with a capital A, which revealed the power of Opinion with a capital O.

The Dreyfus affair and the emergence of modern public opinion

Over the last few decades of the nineteenth century, new patterns of interaction gradually emerged. We have already referred to the fact that

23. See also Geiger, "The Institionalization of Sociological Paradigms," p. 241.

France was the first major European nation to introduce universal (male) suffrage on a lasting basis. Compulsory education had further widened the debate on public issues. The press law of the early 1880s favored wide-ranging freedom of expression (although this was somewhat limited again by the "lois scélérates" of the mid-1890s). Popular newspapers (which began to spring up toward the end of the Second Empire and the beginning of the Third Republic) were flourishing. The largest first hit the million copy mark in the memorable days of the late nineties. The rapid rise in newspaper circulation during these decades preceding the First World War, and its relative stabilization during the decades preceding the Second World War, can easily be discerned from Table 6 (compiled from Frémy 1977).

Whereas there had been an average of less than one newspaper in two households during the early Third Republic, there was an average of more than one newspaper in every household at the outbreak of the First World War. This emergence of a popular press profoundly broadened the horizons of the average citizen. *Le Petit Parisien,* which was soon to surpass *Le Petit Journal* as the largest daily, noted in an editorial in 1893: "To read one's newspaper is to live the universal life, the life of the capital, of all the towns, of all France, the life of all nations. . . . It is thus that in a great country like France, the same thought, at one and the same time, animates the whole population. . . . It is the newspaper which establishes this sublime communion of souls [minds] across distances."[24] The emergence of the popular press, then, created a major precondition for the emergence of modern pubic opinion: not the elite opinion that liberal and other theorists had long postulated as a key to democratic functioning, but a true mass opinion. It communicated new moods to the public of a whole city or an entire country. The political consequences of this new situation were far-reaching.

One of the public moods cultivated by a vast majority of newspapers was nationalism. Some, however, sought to exacerbate it to the extreme, in order to make political capital out of it. The odd alliance of Blanquist socialists, radicals, militarists, monarchists, and clericalists that had lined up behind Boulanger a few years earlier now sought to revive itself around a new unifying theme, and found it in xenophobia.

Paris had gradually become the cultural capital of Europe, and France had become the most cosmopolitan nation of Europe. In his

24. October 13. Quoted by Zeldin, *France 1848–1945*, Vol. 4, p. 181.

The crowd waiting for news outside the building of *La Cocarde*, as seen in an 1890 French book on the press by Clovis Hugues. (From Cl. Bellanger (ed.), *Histoire générale de la presse française*. Paris, Presses Universitaires de France, 1972, opp. p. 288. Original designed by L. Tinayre, engraved by F. Noël, in Cl. Hugues, *Le Journal*, 1890.)

Table 6. *Evolution of newspaper circulation under the Third Republic in France (in thousands)*

Newspapers	1867	1880	1912	1939
Le Temps (1861)		23	45	69
Le Petit Journal (1866)		584	850	178
Le Figaro (1866)		105	36	81
Le Petite République (1875)		196	47	—
Le Petit Parisien (1876)		39	1,295	1,422
La Lanterne (1877)		155	28	—
L'Intransigeant (1880)		72	46	134
La Croix (1883)		—	300	140
Le Matin (1884)		—	647	313
Total Paris	763	2,000	5,500[a]	5,500
Total provinces	200	750	4,000[a]	5,500
Total France	963	2,750	9,500[a]	11,000
Population	37,000	38,000	40,000[a]	41,000
Inhab. per copy	38.9	13.8	4.2[a]	3.7

[a] 1914 data.
Source: D. and M. Frémy, *Quid 1978* (Paris, RTL/Laffont, 1977), p. 1322.

study on *Intellect and Pride* in France, Zeldin (p. 16) gives the figures shown in Table 7.

Around the turn of the century, foreign residents and recently naturalized citizens amounted to some 4 percent of the total population (and at least double that proportion of the active population). Many of these people were immigrant workers from neighboring countries. This created quite a bit of friction with local labor in certain border areas, but they were marginal in the power structure and could therefore hardly be depicted as a direct threat to the nation. On the other hand, the Catholic press made much more out of the Free Masons, Protestants, and Jews in high places in the big cities – who often originated from the "betrayed" Alsace-Lorraine region and had German names at that.

Since the Middle Ages, Jews had often been excluded from the possession of agricultural land throughout Europe and from the practice of some traditional trades. This had often led them to live in the urban areas and enter "modern" professions ranging from finance and trade to intellectual and creative endeavors. In the course of the capitalist, industrial, social, and democratic revolutions, many of them came to play a key role in promoting secular liberalism.

(Top) A newsstand goes up in flames during a demonstration in 1893. (Bottom) The following day, people read the paper next to an overturned newsstand. (From R. de Livois, *Histoire générale de la presse française*, Vol. 2. Lausanne: Spes, 1965, p. 341. Original in Bibliothèque Nationale.)

Table 7. *Foreign residents and naturalizations in France during the latter half of the nineteenth century*

Year	Foreign residents	Naturalizations (previous decade)
1851	380,000	13,500
1861	497,000	15,300
1872	741,000	15,300
1881	1,001,000	77,000
1891	1,130,000	170,700
1901	1,034,000	221,800
Total		513,600

Source: T. Zeldin, *France 1848–1945,* Vol. 3. *Intellect and Pride* (Oxford: Oxford University Press, 1980), p. 16.

Jews became prominent among entrepreneurs and workers, scientists, and artists.[25] Some stuck to orthodoxy, but most assimilated. In spite of this, many remained marked as cultural outsiders to a certain extent. This ambiguous position may help to explain their outstanding contributions to cultural innovations in general, and to the emergence of psychosocial science in particular.[26] Their place "apart" from the Christian "mainstream," however, made it easier to associate them with the rapid changes "threatening" traditional society, to depict them as agents of subversion. It also proved conducive to the revival of antisemitism.

The pogroms of the 1880s and 1890s had accelerated Jewish migration from Eastern to Central Europe, and to a lesser extent to Western Europe.[27] The French Revolution had emancipated the Jews. During the late nineteenth century, however, their number did not exceed

25. An interesting though controversial study of their socioeconomic position in certain European countries *La conception matérialiste de la question juive* was written by a Jewish Trotskyite from Belgium, Abraham Léon. It was reconsidered by Rodinson a.o. in 1971. On the same subject see Ahrendt, *The Origins of Totalitarianism,* ch. 2.
26. Keller, *Und Wurden Zerstreut unter alle Völker* (p. 465), mentions Freud, Marx, Lombroso, Lazarus, and Steinthal, among many others, but omits Durkheim and a great many other early sociologists and psychologists.
27. Léon (*La Conception Matérialiste de la Question Juive,* ch. 7) mentions a migration from Eastern Europe of an average of 3,000 Jews per year between 1800 and 1880, 50,000 per year in the 1880s and 1890s, and then 135,000 per year until the First World War. Lobel (in Rodinson and others) mentions a migration of 3.5 million between 1880 and 1914.

80,000 in France; only a tiny fraction were recent arrivals.[28] Yet, they provided a convenient target for French nationalism, capable of uniting proletarians, shopkeepers, and bourgeois alike against "the enemy within" – against "Rothschild, Ravachol & Company" – as one antisemitic leader put it.[29]

It was suddenly "discovered" that prominent bankers of the controversial Panama canal project (Reinach and Herz), as well as prominent socialist theorists (Marx, Lassalle) were Jews of German extraction. Agitators found their "plot" theory went down extremely well among certain sectors of the population, including the most rebellious ones. The French nationalist leader Barrès said "the anti-Jewish formula" was extremely powerful in mobilizing the population because it was based on hatred. "Listen to the crowd shouting 'Down with the Jews' in the meetings – it is 'Down with social inequality' which one should hear." The French antisemitic leader Drumont confirmed that "the masses . . . more surely guided by their instincts than we are by our knowledge, despise the conservative party, they flee it just like horses flee the dead."[30] At the time, therefore, the German socialist leader Bebel plainly characterized antisemitism as the "socialism of fools."

This foolishness was provided with a certain air of scientific validity by the race theories of the day, however, which identified the Jews as a "parasitic people" living off the knowledge and wealth of the "superior Aryans." In the book preceding his *Psychology of Peoples,* Gustave Le Bon alluded to the Jewish ban on idols, and concluded that

> the Jews have possessed neither arts, nor sciences, nor industry, nor anything of that which constitutes a civilization. They have never made even the faintest contribution to the edification of human knowledge. They have never surpassed that semibarbarous state of peoples without a history. If they ended in having cities, it is because living conditions, amidst neighbors which had arrived at a superior level of evolution, made it a necessity for them.[31]

It is true that the publication of an earlier version of this chapter in the *Revue Scientifique* had elicited immediate criticism. But Le Bon per-

28. Gagnon, *France since 1789,* p. 260.
29. The Marquis de Morès. See Sternhell, *La Droite Révolutionnaire,* p. 178.
30. Sternhell, *La Droite Révolutionnaire,* pp. 209, 197 respectively (my translation).
31. *Les Premières Civilisations,* p. 613 (my translation).

sisted (and the discussion was proudly reprinted only a few years ago by political admirers, as an example of a "courteous exchange").[32]

It should be acknowledged that in those days, many other authors held similar views. That same year, Georges Vacher de Lapouge published his notorious *L'Aryen – Son Rôle Social,* and Edouard Drumont had published his equally notorious *La France Juive* just a few years before. Antisemitic newspapers and organizations were springing up everywhere. They were eagerly waiting for a major opportunity to demonstrate the "treachery" of the Jews. In 1894 they seemed to have found it at last.

During the spring of that year the largest daily, *Le Petit Journal,* started a serialized novel, *Les Deux Frères,* written by Létang. It told of obscure schemers who stole a batch of secret documents from the French War Ministry, added a note manifestly offering them to the enemy, imitated the handwriting of a captain they meant to implicate, had the incriminating materials "discovered" at his home, and then tipped off a newspaper to stir up a public outrage. This scenario may have given some people an idea, because it closely resembled the affair that unfolded later in the autumn of that same year.

The first phase of the affair began when antisemitic officers at the French intelligence agency claimed to have discovered a note offering secret documents among the wastepaper of the German military attaché in Paris. They also claimed to have traced it to a Jewish captain working at the French War Ministry, originating from the contested Alsace-Lorraine region and with a German name: Dreyfus. Experts (including the criminologist Bertillon) later confirmed that the handwriting closely resembled his. The accusers claimed to possess even more incriminating clues, but declined to produce them because of their "confidential nature." They also tipped off Drumont's newly founded antisemitic daily *La Libre Parole,* which had earlier revealed the role of Jewish bankers in the Panama scandal.

On the eve of Christmas 1894, Dreyfus was condemned for high treason. According to one observer, a large crowd outside the prison and courthouse in the Rue du Cherche Midi (that is to say, around the corner from where Tarde lived) greeted the verdict with savage cries of approval, which continued to resound throughout the festive Latin Quarter the whole evening. Shortly after New Year's Day, Dreyfus was

32. *Rôle des Juifs dans la Civilisation* (Paris: Les Amis de Gustave Le Bon, 1985).

Le Petit Journal

SUPPLÉMENT ILLUSTRÉ

Huit pages : CINQ centimes

DIMANCHE 13 JANVIER 1895

LE TRAITRE

Dégradation d'Alfred Dreyfus

"The degradation of the traitor Dreyfus at the École Militaire." Front page of the illustrated supplement of *Le Petit Journal,* Jan. 13, 1895. (From Cl. Bellanger (ed.), *Histoire générale de la presse française*. Paris, Presses Universitaires de France, 1972, opp. p. 289.)

stripped of his rank before the assembled troops at the nearby École Militaire, a lynch mob pressing against the gates.[33] The following month he was sent off to the notorious prison colony of Devil's Island

33. The radical politician Reinach (a relative of the aforementioned banker), in his seven-volume *Histoire de l'Affaire Dreyfus* (Vol. I, chs. 10, 11), quoted these observations from contemporary press reports.

(off Guyana) by a howling crowd. There was just one snag: He kept proclaiming his innocence. His elder brother, a doctor friend (with an interest in hypnotic suggestion, by the way), and a Jewish journalist took up his plight.[34] But it was to take three full years, before their timid protests found a significant echo in public opinion. The second phase started when a new head at the intelligence agency found that secret information was still being passed on to the German embassy. The prime suspect was a dubious major by the name of Esterhazy, and his handwriting more or less matched the original note as well. After an aide came up with more fake evidence against Dreyfus, and after it became clear that his superiors wanted to take him off the case, he confided in a lawyer friend – who in turn alerted the vice-president of the senate, Scheurer-Kestner, himself a Protestant with a German name. By then, others had begun to take an interest in the case as well. Among them was Émile Zola: an anticlericalist (and antimilitarist) of Italian descent, who had just completed his multivolume cycle of best-seller novels, *Les Rougon Macquart*. He contributed various articles to *Le Figaro* denouncing the crowd psychology of hate mongers, and the popular masses' "unconscious" longing for another caesar.[35] When the liberal-conservative daily turned reluctant, he went over to the newly founded radical newspaper *L'Aurore* to continue his crusade.

In mid-January 1898 the course of events accelerated. On the 11th, the real culprit Esterhazy was acquitted by the court, and acclaimed by a joyful crowd. Scheurer-Kestner was not reelected as a vice president of the Senate. The prospects of the Dreyfusards looked dim, unless something dramatic occurred. Zola therefore decided to shift his polemic into a higher gear. He wrote his famous open letter to the President of the Republic, claiming that the army high command had willfully tampered with the evidence, and daring them to sue him. He once again referred to crowds, psychology, and the collective madness induced by the press campaigns.[36] The chief editor, the radical leader Clémenceau himself, converted the oft-repeated phrase "*J'Accuse*" from the end of the article into a headline covering the entire front page. It caused a sensation.

A few days later, *L'Aurore* published a "Manifesto of the Intellectuals," in which a number of well-known scientists and writers joined Zola in his call for a fair trial. It was here that the word "intellectual"

34. Matthieu Dreyfus, Dr. Gilbert, and Bernard Lazare. See Gauthier, "*Dreyfusards,*" p. 14.
35. See Zola, *L'Affaire Dreyfus*, pp. 39 and 41.
36. Zola, *L'Affaire Dreyfus*, pp. 115–22.

CINQ Centimes – Paris et Départements – CINQ Centimes

LA LIBRE PAROLE

La France aux Français!

Directeur : EDOUARD DRUMONT

LE TRAITRE DREYFUS.-LE COMPLOT JUIF

A Newport.-Interview du capitaine et de l'équipage du *Non-Pareil*

LE RÉTABLISSEMENT De la Course

A NEWPORT

Cinq Centimes

L'AURORE

Littéraire, Artistique, Sociale

ERNEST VAUGHAN

J'Accuse...!

LETTRE AU PRÉSIDENT DE LA RÉPUBLIQUE

Par ÉMILE ZOLA

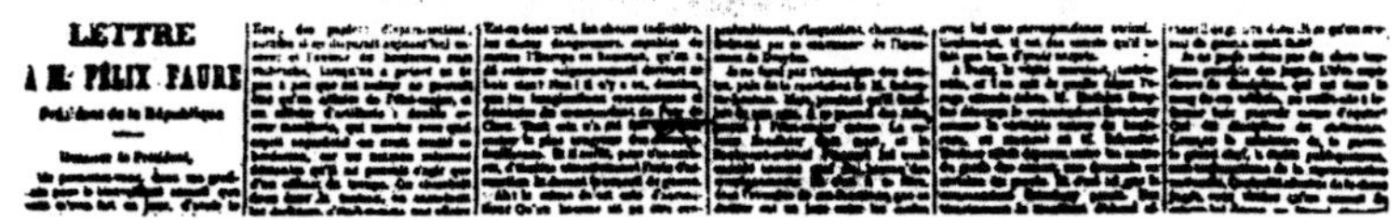

LETTRE A M. FÉLIX FAURE Président de la République

The escalating press campaign. (Top) Drumont's antisemitic daily *La Libre Parole* titled "The traitor Dreyfus – The Jewish plot" (Sept. 10, 1896). (From R. Gauthier (ed.), *"Deyfusards."* Paris: Julliard, 1965, after p. 132.) (Bottom) Clémenceau's radical daily *L'Aurore* titled "I Accuse – Letter to the President of the Republic by Émie Zola" (Jan. 13, 1898). (From W. Keller, *Und wurden zerstreut unter alle Völker*. Munich/Zürich: Knaur, 1966, after p. 272.)

acquired its present-day meaning of "opinion leader." The antisemites reacted with dozens of demonstrations clamoring "Down with the Jews" and "Long Live the Army." They mobilized an estimated four thousand in Marseille, Bordeaux, and Angers; three thousand in Nantes; two thousand in Rouen; more than a thousand in Lyons and Dyon, and crowds also gathered in Reims, Nancy, Saint-Dié, Bar-le-Duc, and many other towns.[37] The demonstrations soon turned into riots attacking Jewish houses. A veritable pogrom occurred in the French colony of Algeria, where some 160 shops were plundered, set afire, or devastated.

The politician Poincaré compared the nationwide agitation to a "nervous breakdown." His radical colleague Reinach later interspersed his famous reconstruction of the events with references to hypnotic suggestion, and even devoted several pages to an extensive summary of the crowd psychology of Sighele, Fournial, Le Bon, and Tarde.[38] Meanwhile, Zola had indeed been sued, and faced the court in February. But the outlook remained grim. That month, the revisionist intellectuals revived the "Ligue (pour la Défense) des Droits de l'Homme," a group totaling some 8,000 people. Facing them, however, was a far more formidable alliance of anti-Dreyfusard leagues. Later that year, for instance, the "Ligue de la Patrie Française" was founded; it came to represent some 100,000 people.[39]

Among the early crowd psychologists themselves, none played a particularly prominent role in the debate. But the psychopathologist Le Bon was clearly an anti-Dreyfusard, who, for once, denounced the cosmopolitan elite and praised the xenophobic masses in his *Psychologie du Socialisme* (pp. 61–2 and 340, among others). By contrast, the leading criminologists of the French school, such as Lacassagne and Tarde (like those of the Italian school, such as Lombroso, Ferri, and Sighele) defended the rule of law. Tarde never took "political" positions in his writings, however, nor is there evidence that he "rushed to sign a petition supporting Dreyfus' cause," as Barrows would have it.[40]

37. Sternhell, *La Droite Révolutionnaire*, p. 232.
38. Poincaré's speech of January 30 in Limoges is quoted in Zevaes, *L'Affaire Dreyfus*, p. 119. Reinach's *Histoire de l'Affaire Dreyfus* references to crowd psychology are particularly elaborate on pages 28–9, but can be found throughout the third (and to a lesser extent the other) volumes.
39. These figures are given by Debray, in *Le Pouvoir Intellectuel en France*, p. 67, n. 1.
40. As Barrows (1981) notes on the last page of her book, in what may have been a slip of the pen.

Durkheim, too, was prudent at first: being "a Jew from the northeast with a German name" himself, and married to a (not closely related) Dreyfus at that. But he soon made up his mind, and became secretary general of the Bordeaux chapter of the "Ligue pour la Défense des Droits de l'Homme." When Brunetière (later an activist in the opposing "Ligue de la Patrie Française") attacked "the intellectuals," Durkheim published an article in the *Revue Bleue*, defending their special role in society.[41]

The dividing line between anti-Dreyfusards and Dreyfusards often ran through institutions, parties, and even families. For two years, it was hardly possible to have a dinner party without people quarreling over The Affair. During that same period, public figures fought no less than twenty well-publicized duels over it.[42]

The third phase was a protracted struggle, in which first one side, then the other, seemed to have the upper hand. For the time being, the revisionist bloc remained fairly small. In the spring of 1898 elections, the Left continued its advance, but several outspoken Dreyfusards were not elected – while fifteen nationalists and four antisemites were. In spite of new revelations on procedural errors committed during the original trial, and a suicide note by an officer involved in forging evidence, Zola felt forced to flee to London in order to escape arrest.

The fourth phase began only in February 1899, when president Félix Faure, who had consistently opposed a revision, died during a secret assignation with a woman. Anti-Dreyfusard leaders (encouraged by monarchist backers) tried to turn his funeral procession into a protest march on the presidential palace and a military putsch, but they failed. The new president proved more open-minded, and by June a new cabinet was formed in which the republican parties put aside their differences in the interest of a united defense of the constitution. Zola returned from exile, and Esterhazy confessed. But the army high command still refused to yield. Upon his retrial, Dreyfus was condemned again, albeit with "extenuating circumstances." He was pardoned shortly thereafter, but it was only seven years later, on the eve of a 14th of July, that he was fully rehabilitated (and given a decoration for his plight). The law had prevailed, but only very, very narrowly. It was another lesson, both for the defenders of parliamentary democracy and its enemies.

41. Lukes (1973/1981), *Émile Durkheim*, ch. 17; Goddijn, *Sociologie, Socialisme en Democratie*, ch. 5.
42. Zevaes, *L'Affaire Dreyfus*, p. 161, n. 1.

(Top) "First of all: let us not speak about the Dreyfus affair!" (Bottom) They did discuss it. Cartoon by Caran d'Ache illustrating the divisive effects of the affair on French public opinion. (From: C. Grimberg and R. Svänström, ***Histoire universelle***, Vol. 11 (Fr. ed). Verviers: Marabout, 1965.)

What was learned? The authoritarian nationalists, who had already discovered the potential of the caesar-type formula during the Boulanger episode, discovered the power of racist agitation during the Dreyfus affair. It exploited suspicion and hate which crowd psychology had identified as more powerful motives than solidarity and love. Furthermore it discovered that the resulting violence could be used to intimidate selected opponents. Ahrendt said: "What happened in

France in the eighties and nineties happened thirty and forty years later in all European nation-states."[43]

The defenders of constitutional rule coined the term "intellectuals" in the modern sense of the word, and discovered their potential as opinion leaders. They also discovered the true nature of public opinion, and the positive role it can play in a mass democracy. The battle of crowds had turned into a battle of publics. On the one hand there was a majority of newspapers headed by the antisemitic *La Libre Parole,* circulating 500,000 copies on its best days. It changed the nature of the lynch mob, urging its public to join the crowds and trying to recruit demonstrators as readers again. On the other hand, there had been a minority of newspapers headed by the Dreyfusard *L'Aurore,* circulating 300,000 copies on its best days. It tried to debunk the general paranoia with a constant appeal to reason and justice.

Halasz, in a book on the affair subtitled *The Story of a Mass Hysteria,* concluded, "The disease which afflicted French public opinion appears as the first instance of its kind in history whose full course was documented in print. The press offered not only a day-by-day recording of the mass delusion and its heroic cure, but was also the medium through which the event itself came to pass. The top figures of the Dreyfus Affair were writers; the events themselves, articles in newspapers; the combat, polemics; the weapon, the pen."[44] An entirely new era had begun. It was precisely this fact, its background and its consequences, that Tarde discussed in his key 1898–99 articles.

Tarde's studies on crowds and publics, conversation and opinion

Tarde's move to the capital completely changed his life. He began frequenting the literary salons of Paris, met many of the intellectual celebrities of his day, and often struck up a correspondence with them. His house at number 62 Rue Saint Placide was right in the heart of the agitated Latin Quarter, halfway between the two geographical poles of the Dreyfus affair: the École Militaire and the Sorbonne. During these years, he clearly became more closely interested in the political pro-

43. Ahrendt, *The Origins of Totalitarianism,* p. 79. The difference was, according to some, that several of the French antisemitic leaders acknowledged that they simply exploited the theme, whereas many of the German nazi leaders actually believed in it.

44. *Captain Dreyfus,* etc., p. 268.

cess. Earlier he had applied his general theory of social change to law, in *Les Transformations du Droit* (1893; the transformations of law). Now he applied it to politics in *Les Transformations du Pouvoir* (1898; the transformations of power). Still later, he was to apply the amended "repetition-opposition-adaptation" framework to economics.[45]

Throughout Tarde's work of the 1890s, the emphasis continued to shift. This is obvious from the gradual change in his use of the key notion of imitation. Nowhere in his voluminous works does he give a precise definition of this process, but he likened it to other topical subjects. In the first edition of *Les Lois de l'Imitation* (1890; the laws of imitation). he likened it to suggestion, for instance. "Society is imitation," he said, "and imitation is a kind of somnambulism."[46] At the time he strongly emphasized the role of leaders (*meneurs*) and hierarchy (imitation from top to bottom, from inside to outside). In his first article on crowd psychology, too, he largely conceived of imitation and suggestion as unilateral processes. In the preface to the second edition of 1895, however, he already likened imitation to "any interspiritual photographic imprint, so to say, whether intentional or not, passive or active."[47] Still later, he even came to speak of interreflection.[48]

The prefix "inter" was rather rare in his early works, but it came to dominate his later thought: interspiritual, intermental, interpsychological. In his last articles on crowd psychology, then, imitation came to mean interaction as a continual and *mutual* process. This also paralleled the gradual widening of his scope: from crowd events proper to social movements and currents of opinion. Whereas social influence in assembled groups may well be conceived as a form of suggestion, he felt, social influence in dispersed groups is better thought of as a form of interaction. By continuing to shift the emphasis, Tarde cut loose from the old paradigms of crowd psychology and made it possible to bypass and transcend Le Bon's limited approach. Recognition of this is often missing from historiographies of the crowd psychologies of the period and puts Tarde's contribution in an entirely different perspective.[49]

45. *La Psychologie Économique* (1902).
46. Tarde (1890/1979), p. 95.
47. Tarde (1890/1979), p. viii.
48. Tarde (1898/1901), p. 29.
49. Both Moscovici's 1981 chapters on Tarde's crowd psychology and Boef's 1984 book on crowd psychology and the interactionist tradition fail to note the link. Lubek's 1981 study on the "lost social psychology of Tarde," by contrast, does recognize its importance.

Tarde's *Études de Psychologie Sociale* (1898) was one of the very first books in any major Western language to include that term in their title. That year, Tarde "discovered" the nature and power of modern public opinion. This coincided with the height of the Dreyfus affair. When I interviewed Tarde's youngest son Guillaume in 1981, he confirmed that it was highly plausible there was a link, and referred back to that memorable day in January 1898 when as a thirteen-year-old boy he heard newspaper vendors in the nearby Rue de Rennes shout the headline of Zola's sensational attack on the military high command in the daily *L'Aurore*. He remembered how his father sent the maid for a copy, and after reading it exclaimed repeatedly: "C'est extraordinaire!" The family also maintains that Tarde was an outspoken Dreyfusard, and got embroiled with various colleagues over the affair.[50]

Tarde's key article "Le Public et la Foule" was published in the *Revue de Paris* in July of that year, just when Zola felt forced to flee abroad. The article contained only two explicit references to the Dreyfus affair, and two more to Drumont and antisemitism. Upon close reading, however, one can easily discern a dozen more implicit allusions to the affair, particularly in the opening (pp. 5, 12, 13, 14, 15, 17) and concluding sections (pp. 52, 57, 58, 59, 60, 61). From the last section, one can also infer that it may have been Tarde's original intention to write an article about "the crimes of publics," just as he had written earlier on "the crimes of sects" and "the crimes of crowds." But he apparently changed his mind in the process, recognizing that the whole question had much wider implications.

As observed in the introduction to this chapter, Tarde began the article by contrasting publics and crowds. Crowds, he said, were a primitive, natural phenomenon determined by "the action of the looks of others." Publics, on the other hand, were a civilized, cultural phenomenon determined by "the thought of the looks of others." "One can be part of only one crowd at a time. From this follows the far greater intolerance of crowds, and consequently of nations dominated by the spirit of crowds." By contrast, "one can belong – and in fact one always does belong – simultaneously to several publics." Thus it follows that "the gradual substitution of publics for crowds . . . is always accompanied by progress in tolerance" (Tarde in Clark 1969, p. 281). Yet, this difference should not be overestimated. "The man of one book is to be feared, it has been said; but what is he beside the man

50. Interview with Guillaume de Tarde in his Paris appartment, November 18, 1981.

of one newspaper! This man is each of us at heart, or nearly so, and therein lies the danger of modern times" (p. 283).

The advances in communication created the "sensation of actuality," Tarde maintained, and the improvement of printing techniques involved ever wider groups in opinion formation. Popular newspapers consolidated political organizations, and generated currents of opinion. Meanwhile there was a high degree of mutual penetration among various publics, resulting in a larger (Public) Opinion. Whereas physical and biological factors (race, sex, age) still played a major role in crowds, psychological and social factors (faith, goals) dominated publics. Both crowds and publics had four "forms of existence": expectant, attentive, expressive, or active. (Note that these terms have occasionally been mistranslated in English.)[51]

All in all, Tarde continued, publics are less cruel than crowds, although crowds are always inspired by publics. Furthermore, individuals are always superior to both publics and crowds. "The danger for new democracies is the growing difficulty for thoughtful men to escape the obsession and fascination of turmoil." The contribution of "intellectuals," therefore, lay not so much in their discoveries, he concluded, but in "their force of resistance. Let them beware if they should separate!"

His other "crowd" article of this period, "L'Opinion et la Conversation," extended this argument. It was published in the same *Revue de Paris* in late August 1899, immediately after Zola's return from exile and Dreyfus's retrial. A manuscript of the article is in the Lacassagne archives in Lyons. The title page bears a large handwritten comment: "Maybe add *some pages on the Affair???*" (emphasis in the original).[52]

Tarde probably decided against it: There were only a few minor additions. The article once again contained a number of indirect references to the affair, but no direct discussion of it. Tarde may have felt that in the eyes of some, his scientific reflections would inevitably be turned into a political polemic. He may even have been only partly aware of his major source of inspiration. Yet his discussion of public

51. Tarde (1901, p. 38) speaks of "manières d'être." The American edition (Clark 1969, p. 290) wrongly translates this as "stages of being." The second point is less obvious, but in view of the elaboration Tarde gives, I would prefer "expressive" instead of "demonstrating" as a translation for "manifestantes" in his categorization of crowds (compare pp. 42–3 in the French edition with pp. 292–3 in the American one). Also see the remarks on Park and Blumer in the last section of this chapter.

52. Ajouter peut-être *quelques pages sur l'Affaire???*

opinion and its roots in "informal communication" was clearly linked to the events of the day.

The article discussed the transformation of conversation from village gossip to the literary salons of the Enlightenment period, noting the simultaneous expansion of letter writing and postal services.[53] "The newspaper has thus finished the age-old work that conversation began, that correspondence extended, but that always remained in a state of sparse and scattered outline – the fusion of personal opinions into local opinions, and this into national and *world* opinion, the grandiose unification of the public mind" (Tarde in Clark 1969, p. 318). Thus emerged modern public opinion. "Opinion, as we define it, is a momentary, more or less logical cluster of judgments which, responding to current problems, is reproduced many times over in people of the same country, at the same time, in the same society" (p. 300).

The emergence of public opinion profoundly changed the nature of politics: "It suppressed the conditions which make possible the absolute power of the governing group. This power was greatly favored, in actuality, by the local splitting of opinion" (p. 302). "Universal suffrage and the omnipotence of parliamentary majorities were only made possible by the prolonged and accumulated action of the press, the sine qua non of a great leveling democracy" (p. 305). Since the Enlightenment, therefore, people had been interested in "the study of interspiritual relations," "interpsychology," and "social psychology," Tarde claimed (1899 in Tarde, 1901, p. 140). He further emphasized this point in his preface to the book edition of his articles, *L'Opinion et la Foule* (1901). As we have seen in the introduction to this chapter, it advocated a "collective or social psychology" based on the study of "the mutual relations between minds" (p. v).

In 1900–1, Tarde also published other articles, "L'Esprit des Groupes" and "La Psychologie Intermentale."[54] At the Collège de France, he lectured on "intermental psychology" during his first year, and on "interpsychology" during his last year. Some of his notes on these subjects survive today. An unpublished manuscript on interpsychology outlined it as "the science of the relationship between

53. According to Tarde, the number of letters sent in France rose from 2.5 million in 1700 to 10 million in 1777, 63 million in 1830, and 773 million in 1892 (Tarde 1969, p. 315). Note that the modern meaning of the word "public" also dates from the Enlightenment. (See the *Larousse de la Langue Française*, p. 1452.)

54. In the *Archives d'Anthropologie Criminelle*, 15:5 ff., and the *Revue Internationale de Sociologie*, 9:1–13, respectively.

[various] consciousnesses." It was to be distinguished from both psychology (which Tarde said was "devoted to consciousnesses taken in their individuality") and sociology ("which it explains but does not constitute").

At one point in the document he said "Interpsychology studies the behavior [!] of groups." At another point he spoke of the "interactions [!] between consciousnesses," and specified these as those between individuals, and those between individuals and groups (such as crowds and publics). He also identified the three "technical procedures" to be employed: (1) the creation of laboratories; (2) the use of statistics; and (3) direct observation.[55] One of the projects Tarde considered in this context was the direct observation of group behavior in schoolyards. It is clear, then, that Tarde was really on the brink of elaborating a surprisingly modern social psychology when he died in 1904.

Tarde and the American connection

In the first few years after his death, Tarde's sons attempted to continue his work; particularly the second son, Alfred de Tarde (1880–1925). According to the key study by Lubek (1981) previously mentioned, his law thesis elaborated certain concepts from his father's last major work, *La Psychologie Économique*. Together with Teutsch, he founded a *Revue de Psychologie Sociale* as early as 1907, later renamed *La Vie Contemporaine*, but it disappeared in 1908. Together with Massis (under their common pen name Agathon), he also carried out one of the first primitive polls in France, published in the journal *L'Opinion* (later to become *La Vie Française*). They did a number of interviews (though not standardized) with a sample (though not representative) of students from *lycées* and *grandes écoles*, in order to determine "the mood" of the future elite. They also had some well-known figures react to its results. In line with their own neoconservative sympathies, they observed a "taste for action, a new political realism, patriotic faith, and a revival of Catholicism."[56]

In 1909 Tarde's sons also edited a short biography and an anthology introduced by Tarde's successor at the Collège de France, the philoso-

55. See Milet (1970), pp. 396–9 (my translation). Also see Tarde (1973), pp. 169 ff.

56. See *Les Jeunes Gens d'Aujourd'hui* (1913), which attracted so much attention that it had to be reprinted many times during that same year. Alfred de Tarde also defended his father's legacy with heavy and sometimes dubious attacks against Durkheim.

pher Bergson. That same year, a monument was inaugurated in Tarde's native Sarlat. A year later, Matagrin published an elaborate study on *La Psychologie Sociale de Gabriel Tarde*. After that, however, Tarde's influence began to wane in France. Chapters were still devoted to Tarde and/or his interpsychology in handbooks such as Dumas's *Traité de Psychologie* (1924, vol. 2, book III, ch. 3), Blondel's *Introduction à la Psychologie Collective* (1928, first part, ch. 3), and Essertier's *Philosophes et Savants Français au XXe Siècle* (1930, vol. 5, ch. 3). But a Tardean tradition as such never really took off, and when opinion research and social psychology finally emerged in France after the Second World War, they were largely imported from abroad. Curiously enough, many of the French pioneers in these fields were unaware of the true extent of Tarde's contributions – even to certain foreign approaches that they were now ready to embrace. To illustrate this, I will cite one key figure from two successive postwar generations of social psychologists.

The first is Jean Stoetzel, the founder of modern opinion research in France. He did mention Tarde as a pioneer of "group psychology" in his doctoral thesis *Esquisse d'une Théorie des Opinions* (1943; outline of a theory of opinion). But he failed to mention him as a pioneer of opinion theory in many of his subsequent works. Stoetzel's contribution to Gurvitch's *Traité de Sociologie* even went so far as to deal with "the psychology of interpersonal relations" without referring to Tarde and also failed to include him in a list of founders of social psychology.[57]

The second key figure is Serge Moscovici – one of the most prominent present-day social psychologists in France, who also published several studies on collective behavior. In his interesting book on the social role of active minorities, he criticized dominant paradigms and proposed an alternative, "genetic" model without apparently being aware that it had various elements in common with that of Tarde's American successors of the Chicago School (the significance of which we will return to). His more recent study on crowds emphasized the role of Tarde as a kind of *trait d'union* between Le Bon and Freud, but failed to mention his contributions to the interactionist tradition and its

57. Milet (1970), p. 55; Lubek (1981), p. 372. See also Jean Maisonneuve, "Naissance et développement d'une discipline en France – La psychologie sociale" in R. Boudon, F. Bourricaud, and A. Girard, eds. (1981); and Jean Maisonneuve, *La Psychologie Sociale*.

alternative approach to crowd psychology.[58] One of the very few social scientists in France to recognize Tarde's originality at an early stage was the statistician Raymon Boudon.[59] Apart from that, Tarde's reputation in France survived only in criminology.[60]

One should of course bear in mind that even during his life, Tarde's social psychology was appreciated by only a small circle of French colleagues. Lubek says the sales of his most famous work in this field, *Les Lois de l'Imitation,* did not exceed 4,600 copies over the first few decades; and the sales of *Les Lois Sociales* did not even exceed 2,500 copies.[61] Yet, several of these copies did travel to French-reading colleagues abroad. Some of Tarde's work was even translated in various foreign languages: Russian, Spanish, German, and English. I have already mentioned that Tarde's thought differed substantially from that of most of his French contemporaries and fit in better with Anglo-Saxon traditions. It may not be surprising, then, that aspects of his approach survived better in the United Kingdom and best in the United States. In fact, his influence on early American psychosocial science was considerable indeed, and in some respects greater than is often acknowledged.

We may begin with his resonance in the field of psychology. Founder of the field in America, William James, later wrote to his British colleague Wallas, for instance: "I myself see things à la Tarde" (quoted in the latter's book *The Great Society,* ch. 8). An even closer link existed, however, to the thought of another major founder of American psychology: James Mark Baldwin (1861–1934). He studied in Berlin and Leipzig, translated Ribot's French book on German psychology, and visited Paris to interview the protagonists of the Salpêtrière-Nancy debate. Upon his return to North America, he founded the first psychological laboratory in Canada, published one of the first psychological handbooks in the United States, and started the *Psychological Review* and other major journals (with Cattell). At

58. Neither his *Social Influence and Social Change* (1976; *Psychologie des Minorités Actives,* 1979) nor his *L'Age des Foules* (1981, *The Age of the Crowd,* 1985) mention Tarde's contribution to the interactionist approach in crowd psychology. The latter book mentions Park and his German thesis (p. 83), but only in relation to Le Bon.
59. See, for instance, Boudon in Tarde (1890/1979).
60. In 1972, the Comité de Coordination des Recherches Criminologiques instituted an annual Gabriel Tarde award.
61. For 1890–1921 and 1895–1913, respectively. See Lubek (1981), p. 367.

Princeton, Baldwin developed major theories on *Mental Development in the Child and the Race* (1894) and *Social and Ethical Interpretations in Mental Development* (1897). The latter book was subtitled *A Study in Social Psychology,* and referred to "the lack in English of a book on Social Psychology which can be used in the universities" (p. vii). In both works, "imitation" was identified as a fundamental social psychological phenomenon.

In the preface to the latter book Baldwin also observed, "I find my opinions in the matter of the social function of imitation lying near those of M. G. Tarde" (p. viii). A year earlier, Baldwin had hailed the "brilliant formulations" of "one of the most authoritative and distinguished writers in sociology and social psychology" in his foreword to the American translation of Tarde's book on social laws. And a year later, he wrote to his French colleague proposing to label imitation "in order of priority" the sociological principle (of) Tarde–Baldwin, since they had both discovered it "independently." At the same time, however, he demanded that the psychological principle be named after himself alone.

But the American edition of Tarde's original study on imitation was delayed that year by problems over payment of the translator, one Park (it is unclear whether he was related to the sociologist of that name, to whose crucial role we will return). The postponement lasted until 1899, when a volunteer resumed the translation, and it was not until 1903 that it came out. Here again, sales probably did not exceed 2,000 copies over the first few decades – and it reached only a limited public of specialists.[62] But over the following years, secondary studies such as Michael Davis's doctoral dissertation *Gabriel Tarde – An Essay in Sociological Theory* (1906) and his book *Psychological Interpretations of Society* (1909) contributed greatly to the spreading of Tarde's fame.

If we look at Tarde's role in the emergence of Anglo-Saxon social psychology, it is interesting to consider the publication of McDougall's *Introduction to Social Psychology* and Ross's *Social Psychology* in 1908 – which put the new interdiscipline on the map and on the agenda in the Anglo-American world. The circulation of both books was far greater than that of all of Tarde's works in this field. The former title sold 100,000 copies over the next few decades, the latter 43,000 copies.[63] But Tarde's mark was clearly visible on both.

62. Lubek (1981), pp. 381–82, 367, respectively.
63. Reported in 1949 and 1936, respectively. See Lubek (1981), p. 367.

William McDougall, the British physiologist and psychologist, who was to emigrate to America later in his career, still remained caught somewhat in the older biological paradigms. His 1908 book focused on the "social instincts" of man, but still devoted an entire chapter to "imitation, play and habit," and referred to Tarde repeatedly. Edward Ross (1866–1951), by contrast, an American sociologist who had traveled to Europe and France, was much more of a Tardean. He concluded the preface to his 1908 book by paying "heartfelt homage to the genius of Gabriel Tarde" and "his incomparable" *Lois de l'Imitation*. According to Clark, Ross's book probably served as the most important vehicle for the diffusion of Tarde's thought in America.[64] Moreover, Ross's earlier work, *Social Control* (1901), was said to follow Tarde's original thought so closely that at times it seemed like a free translation of fragments of his book on imitation.[65]

If we look at Tarde's influence on early sociology in America, finally, we find that there was hardly a major founder of that field who was not familiar with at least some of Tarde's ideas on imitation, and welcomed them as an important contribution. Clark gives a sampling of their opinions: Franz Boas said he was "profoundly impressed" by the book on imitation; Charles A. Ellwood devoted part of his doctoral dissertation to Tarde; Franklin H. Giddings said sociology "owed more" to him than to any other contemporary; William Ogburn was clearly influenced by Tarde; Lester Ward called him "one of the leading thinkers of our time," and so forth.[66] In my opinion, however, the most substantial yet somewhat neglected link was the one between Tarde and the "psychological" sociology of the Chicago School.

On the occasion of the fourth centenary of Columbus's "discovery" of America and the Chicago Universal Exposition, John D. Rockefeller had generously funded the establishment of a university there, which was to lure "the best and the brightest" away from the Ivy League to that booming city. The Sociology Department was headed by Albion Small, who called Tarde "perhaps the most prominent figure just at present among the founders of the new science."[67] It was here that the interactionist tradition evolved. Some of it went back to Charles Horton Cooley, who appreciated Tarde's work because he

64. Ross (1908/1925), p. VII; and Clark (1969), pp. 65–6. Professor Brouwer pointed out to me that Ross visited Paris during the same years of the Dreyfus Affair.
65. House (1936), quoted by Lubek (1981), p. 381.
66. All quoted by Clark (1969), pp. 65–7.
67. Clark (1969).

"emphasized the social psychological foundations of society."[68] Chicago theorist George Herbert Mead, too, owed "some of the focus of his work" to Tarde, according to a specialist.[69]

William I. Thomas and Florian Znaniecki introduced the concept of "attitude" here, in their famous 1918 study, *The Polish Peasant.* Speaking of the diffusion of cultural traits, the latter recognized that "a very consistent theory of this process was developed by Gabriel Tarde."[70] One of the first attempts to "measure" attitudes and social distance was undertaken by the psychologist Emory S. Bogardus, who received his Ph.D. in Chicago and considered Tarde "the chief founder of social psychology."[71] The idea of measurement had been suggested to him by the Chicago sociologist Robert Park (1864–1944), whose role in the diffusion of Tardean ideas has been recognized but underestimated.

The famous *Introduction to the Science of Sociology,* which Park published with Burgess in 1921, had a strong Tardean streak in it. Although it failed to include an excerpt from his work, it contained more references to the French author than to such luminaries as Comte, Durkheim, Simmel, Weber – or even to Cooley and Thomas.[72] Whereas Tarde's entire approach focused on social imitation, Park's approach centered on social interaction (ch. 5). Whereas Tarde assigned a central role to opposition and adaptation, Park assigned it to competion/conflict and accommodation/assimilation (chs. 6–9). This perspective was further elaborated by Herbert Blumer, the founder of the so-called symbolic interactionist tradition.

All of this is important in the context of this study in that it reveals a strong and direct link between Tarde's latter-day work and thought, and a major approach to collective behavior today – a link that has been overlooked by the other historians in this field.

As mentioned in the introduction to this study, two major traditions stand out today. On the one hand the social psychological "deindividuation" approach, first outlined by Festinger and his like, and later elaborated by Zimbardo and others. It focuses on a form of psychological regression and on a supposed reduction of the critical

68. Coser, *Masters of Sociological Thought,* p. 322.
69. Strauss in Mead (1956/1977), p. xxviii.
70. Znaniecki quoted by Grupp (1968), p. 344.
71. Bogardus (1940/1948), p. 395.
72. According to Everett Hughes (1961), quoted by Clark (1969), p. 68, n. 150.

distance between self and others within a group. Its explanatory power is greatest for physically assembled crowds proper (and the behavior of people otherwise anonymous). This theory, one could say, is to a large extent a contemporary offshoot of Le Bon's ideas on hypnotic suggestion. On the other hand, there is also the psychosociological interactionist approach – first outlined by Park and by Blumer, and later elaborated by Turner and Killian, Lang and Lang, and others. It focuses on forms of sociological change and the "natural history" of interaction patterns within a group. Its explanatory power is greatest for physically dispersed movements and currents and for public opinion, attitude change, and mass communication research. This theory, by contrast, may be considered a contemporary offshoot of Tarde's ideas on imitation.

Both psychologist-historians of the Roman or Latin school of collective behavior such as Moscovici, and sociologist-historians of the Chicago approach such as Boef, have tended to neglect this link. It is my contention that it is highly significant. The first link in the chain is Park's German doctoral dissertation *Masse und Publikum* (1904), which did not become available to the English-speaking reader until a lifetime later, as *The Crowd and the Public* (1972). At that time, most readers were unaware that this was almost the same title as Tarde's third essay – identified in this chapter as a turning point in his thought on the subject. Even Ellsner's introduction to the belated translation misses this point.[73] Yet Park's dissertation is partly an extension of Tarde's argument in that essay (alongside with elements taken from Simmel and other continental thinkers).

The same even applies to parts of Park and Burgess's famous *Introduction to the Science of Sociology*. Whereas Tarde had emphasized "the action" and the thought or "the looks of others" in his social and collective psychology, Park maintained that "The mere fact that [people] are aware of one another's presence sets up a lively exchange of influences, and the behavior that ensues is both social and collective" (p. 38). In this sense, he confirmed, crowds are similar to sects and publics (ch. 11, introduction). Whereas Tarde made "collective meditation" one of the key processes, Park did the same with all sorts of social unrest – which resulted in a new "collective definition" of the situation, he said. As a matter of fact, Park's entire *Introduction* culminated in the contrast between social control and collective behavior.

So did Blumer's theory, as formulated in the (second) part on collec-

73. Elsner's later introduction to the American edition, f.i., fails to mention Tarde.

tive behavior in *Principles of Sociology*, edited by McClung Lee (1939). He did not cite Tarde, but his theory contained elements remarkably similar to Tarde's. Like Tarde, he distinguished four (albeit partly different) types of crowds: casual, conventionalized, acting, and expressive (pp. 78–9). The same approach permeates the well-known postwar handbooks of Blumer's pupils Turner and Killian (*Collective Behavior*) and of Lang and Lang (*Collective Dynamics*), both quoted in the introduction to this study.[74] Like Tarde, they all focused on the emergence of new patterns of interaction, and they extended their arguments from crowd phenomena proper to social movements and currents of opinion.

Even the attitude research and opinion research that Europe imported from America during the postwar years contains strong traces of Tardean thought. The modern two- or three-factor theory of cognitive and conative aspects of attitudes was preceded by Tarde's distinction between beliefs and desires. "Cognitive dissonance" theory was preceded by his discussion of their possible interference (see Tarde 1890, pp. 28 ff.). The possibility of attitude scaling was discussed as early as his first article on the subject, which spoke of "the measurability of individual belief and desire" (1880, in Tarde 1969, p. 203). Ten years later he had even stated bluntly: "Only a psychological statistic, reporting on changes in the specific beliefs and desires of individual people – if this were at all possible – would provide the deeper reasons for the ordinary statistical figures."[75]

Modern theories of attitude change and communication effects were preceded by Tarde's emphasis on their mediation by informal communication in conversation and correspondence. The "personal influence" and "two-step flow" theories were preceded by Tarde's emphasis on hierarchy in imitation. According to Clark, he also "came close to outlining a program for content analysis" (in Tarde 1969, p. 44). The reedition of some of Tarde's works beginning in the late 1960s, therefore, and the revival of interest for his thought in both France and America, seems fully justified.[76] And so does the characterization in the mid-1970s' edition of a leading encyclopedia, which called him "one of the most versatile social scientists of his time."[77]

74. Although they also fail to notice the significance of the link. See Turner and Killian, *Collective Behavior* (1957, pp. 4–5; 1972, pp. 13–14); Lang and Lang, *Collective Dynamics* (1961, pp. 210–12 among others).
75. Quoted by Lazarsfeld in Woolf, *Quantification*, p. 199 n. 104.
76. See Grupp (1968), Clark (1969), Milet (1970), Rocheblave-Spenlé in Tarde (1973), Lubek (1981), and others.
77. *Encyclopaedia Britannica*, Vol. IX, p. 824.

Summary and conclusions

This study examined the emergence of crowd psychology and related fields in late nineteenth-century Europe. It focused on the background, nature, and impact of five pioneer texts. Taine's multivolume study on the French Revolution and *Les Origines de la France Contemporaine* (1875–93) was not on crowd psychology proper, but served as a major source of inspiration for all four subsequent authors. The first monograph on the subject as such was Sighele's *La Folla Delinquente* (1891), although its contribution is often misunderstood. The second monograph was Fournial's *Essai sur la Psychologie des Foules* (1892), which is overlooked in most of the literature, but proved to be a major missing link between the Italian and French authors. The best-known book is *Psychologie des Foules* (1895) by Le Bon, who is often mistaken for the founder of the field, but who did indeed exert by far the greatest influence. Another pioneer is Tarde, who published four essays on the subject in 1892–93 and 1898–99; the latter three were collected in his book *L'Opinion et la Foule* (1901). These early major authors, and several related minor ones, are often subsumed under the label of the Roman or Latin school of early crowd psychology.

Each chapter of this study was devoted to one of these five texts and its author. This is not because I favor a "Great Books" approach to the history of science – quite the contrary. In order to be able to dissect the context or even the conjuncture of lines of influence at which each specific view stands, we need a focal point: a time and a place, a text, and an author. The idea is not that these men were lonely heroes who developed new approaches out of the blue. The idea is that their work was embedded in a steady stream of more or less related approaches, underwent and exerted both social and intellectual influences.

Thus each chapter contains at least four elements. A first element is the social context, with emphasis on the political situation and the role of crowd events therein. This relates to the general question of what concrete problems the author and his group were trying to help solve. A second element is the author's life and work: the author's social position and professional career. This also relates to the general question of the institutional setting in which these problems were considered. A third element was the intellectual context: the discipline(s) in which the author was active and the major debates of those days within these same disciplines. This relates to the general question of the authors' fundamental perspectives and conceptual tools. A fourth element was the impact of their crowd psychology and related theories on academic science and social ideas. The question is also whether they meant to address the academic forum only or the wider public as well. Only in view of these four elements did we assess the nature of their main works and their contribution to the evolution of new approaches to the field.

In this last chapter, therefore, we will review the sudden emergence of crowd psychology in the course of the (first half of the) 1890s against these four different backgrounds. In doing so, I will alternate between a long-term and short-term perspective. That is to say, we will consider the evolution of social problems, institutional settings, intellectual debates, and the publics addressed over the previous century and over the previous decade or so. Both seem to be of relevance.

Let us first consider the concrete problems the authors and their groups were trying to help solve. For all five, the initial concern was mob events, although that often was only the starting point of their reflections. Mob events had gradually become a major object of fascination over the preceding hundred years, roughly marked by the year of the French Revolution, 1789, and 1888, the year in which Guy de Maupassant expressed his horror at the sight of even a joyful crowd in his diary *Sur l'Eau*. Mob events played a prominent role in stories, novels, and plays by such famous national authors as Charlotte Brontë, Conscience, Couperus, Dickens, Eliot, Flaubert, Hauptmann, Hugo, Manzoni, Schnitzler, Scott, Stendhal, Tolstoi, and Verga (analyzed in a separate paper).

Although mob events invariably played a major role in all great upheavals of these hundred years, it cannot be said that they really became more frequent or dramatic. It was not so much their nature that changed, but their meaning. This change was related to the economic, technical, social, and political revolution that had taken place. Mob events had become major challenges to the newly established capitalist-industrial and bourgeois-democratic order.

This is even clearer if we limit our scope to the decades in question. After Germany and Italy were each reunited and joined Britain and France as major European powers, the 1870s were still overshadowed by war and revolution, and by the settling in of the new regimes. Though the 1880s were characterized by considerable national and social agitation, it looked as if this opposition could and would gradually be absorbed through continuing progress. The 1890s, however, saw a resurgence of major crises and widening fears of another violent upheaval. This refocused attention on the dangers of the mob. Thus the early crowd psychologists all began by addressing the central question of how crowd excesses occurred, and how they could or should be prevented or repressed.

Sighele, his teachers, colleagues, and students at first reflected primarily on the sporadic unrest of the 1880s. Although they denounced crowd excesses, they often exonerated the people involved. In the course of the 1890s, however, challenges to the established order became more forceful, and Taine's frightful image of the revolutionary mob gained new meaning. Whereas Fournial had been relatively bland, Le Bon's observations on the ease with which Paris crowds could be manipulated were decidedly alarmist. Tarde, too, was rather pessimistic about crowds at first, but developed more optimistic views of publics in later years. Thus for all five authors, contemporary crowd events were a major source of concern.

This brings us to the second question: the institutional setting in which these problems were considered. Here the prime element of change was political organization of the state and of its citizens. Over the previous century, the state apparatus and its forces of repression had expanded greatly. Conscription was introduced, standing armies were maintained to defend the political regime at home or to conquer new territories abroad. Police surveillance had become permanent and widespread, criminals were now systematically tracked down and put away. Courts and lawyers applied legal rules of increasing complexity.

Over the previous decade, both criminology and political science had begun to draw on the existing sciences of man: medicine, anthropology, psychopathology.

If we look at the life and work of the early crowd psychologists in this perspective, we see that both their family background and their professional careers strongly linked them to a "law-and-order" perspective in one way or another. Taine's father was a lawyer; he himself became involved in the founding of a political science school that would help stabilize the republic. Sighele's father was a judge, and he himself a lawyer who became involved in politics, albeit largely oppositional. Fournial was an army doctor involved in colonial explorations, and so was Le Bon at various times. He came from a family of military officers and civil servants, and aspired to become a senior adviser to republican statesmen. Tarde, finally, was the son of a judge and a judge himself, before becoming a top civil servant at the Ministry of Justice. The same could be said of several minor crowd psychologists of those days.

Three of the four major early crowd theorists proper (namely Sighele, Fournial, and Tarde) belonged to the same new small world of army and legal medicine, criminal anthropology, and criminal sociology. The fourth (namely Le Bon) did not belong to it himself but had close links to adjacent fields. Although expert knowledge in all these fields was in growing demand, and new chairs were created throughout the 1880s, however, none of the early crowd psychologists came to occupy a really influential academic position. Only Tarde briefly held a professorship at the prestigious Collège de France, but this was toward the end of his life.

It was not only the organization of the state, however, that changed the meaning of the crowd problem, but also the organization of its citizens. Large-scale voluntary associations sprang up during these years, posing a growing threat to the established order. The gradual extension of civil and voting rights had shifted power away from committees of notables and toward mass parties, leagues, and unions. The gradual improvement of transport and communication greatly expanded their potential for mass mobilization during these same years. Unprecedented crowds could be mobilized at any moment and directed against any target as well. This was demonstrated in the late 1880s and early 1890s by new mass movements of both the Left and the Right.

Nationalism, which often originated in resistance against foreign domination, was gradually transformed into an urge for territorial expansion, whereas xenophobia often merged with racism. Socialism, furthermore, which had long remained a fairly marginal phenomenon, quickly gained in popularity after the founding of the Second International in 1889 and the beginning of the May Day demonstrations in 1890. If any particular series of events triggered a growing interest in crowd psychology during the (first half of the) 1890s, it was this: the growing number of workers going on strike and taking to the streets of all the larger cities of Europe on the same day every year, and with the same demands.

The frequent suggestion that the early crowd psychologists held entirely identical feelings about these events, however, is clearly wrong. We have seen that Sighele's original inspirator was a socialist, whereas Sighele himself can probably best be described as a center-leftist at the time. Fournial's political inclinations were to the Right of the spectrum. Tarde may probably best be described as a center-rightist. Both Taine and Le Bon were rabid antisocialists and clearly held conservative if not reactionary views. To make things more complicated, Sighele's mentor later became a fascist. Yet in spite of their widely diverging positions on a Left–Right dimension, these early crowd psychologists also held certain ideological premises in common.

They all saw themselves as the standard bearers of balanced judgment and scientific rationality. Insofar as they were involved in practical politics, they felt they should help steer the masses away from spontaneous outbursts and "irrational" behavior, and channel their energies into permanent organizations and "constructive" activities. Their social position, too, contained similar elements.

Agewise, they fell into two categories. The most well known and influential were already middle-aged when they published their studies on crowd psychology. Taine was between 47 and 65, Le Bon was 54, Tarde was between 49 and 56. The lesser known and not so influential, by contrast, were only just completing their university education and embarking on a professional career: Sighele was 23 and Fournial 26. Yet they, too, identified with the perspective of mature citizens, denouncing the excesses of crowds as immature, puerile, and childish. All of them being men, they furthermore tended to identify crowds with women. As white Europeans, they tended to identify crowds with "primitive savages." As mostly members of the upper or middle class, finally, they tended to identify crowds with the lower classes. On the whole, then, they saw themselves as rational individuals par excellence, and crowds as their natural antipodes.

The third question is how they conceptualized the contrast between ordinary and exceptional behavior: What theoretical tools they employed to explain crowd behavior. In a long-term perspective, the previous hundred years had been characterized by the growth of science. The early crowd theorists had all received university training. Those were the days of the triumph of positivism: the quest for "hard" facts and the discovery of "universal" laws. Those were also the days of the triumph of evolutionism with its emphasis on physiological differences and adaptation levels. They were lawyers debating questions of free will, moral guilt, and legal responsibility. They were also physicians discussing questions of mental illness and epidemic contagion. In doing so, they used metaphors derived from the discovering of those days: They compared collective arousal to the spreading of viruses and bacteria, to electric and magnetic radiation.

In a short-term perspective, however, the new sciences of individual and society had sprung up beside them. The 1880s had been characterized by the emergence of psychology as an academic discipline, and by debates about normal functions and pathological deviations, about personality characteristics and social influence. Sociology was about to follow with discussions about group contrasts.

Although some founders of these disciplines used experimental research and statistical analyses, however, there was by no means a consensus over correct procedures. The lines of specialization were still vague, and conceptualization was still tentative. Thus the early crowd theorists were really active in different fields, and their work was marked by different paradigmatic debates of those days. Taine was a historian who tried to incorporate the latest philological trend: a critical study of sources and an "objective" reconstruction of events. Yet at the same time he was a psychologist trying to illustrate certain mental mechanisms. Most of all, he was trying to prove the importance of national character, and the dangers of the revolutionary mob. Under its influence, he said, a regression took place, which made people slide backward along the ladder of civilization to the level of savages or even animals.

Sighele, in turn, was a lawyer closely involved in the offensive of the new "positive" school of criminology against the so-called classical tradition. He was interested in complicity as an aggravating or extenuating circumstance. He felt that criminal crowds might exert such a compelling influence on honest individuals that they could only be held semiresponsible for deeds committed under those circumstances.

Fournial was in turn a medical doctor, who thrived on the debates in physical anthropology over brain functioning and mental dynamics. He therefore further elaborated these ideas in a psychopathological sense. In doing so, he prepared the ground for Le Bon, who was also a medical doctor and streamlined these theories in accordance with two major new paradigms. One was the evolution/dissolution model of the Paris school, the other was the hypnotic suggestion model that arose out of the Salpêtrière–Nancy debate. Tarde, finally, was also a lawyer by training, but gradually became interested in elaborating a psycho-sociology or social psychology with a much wider relevance. His famous debate with Durkheim over the relation between psychology and sociology pushed him to elaborate his crowd theories in an interactionist sense. Thus every one of these early crowd psychologists laid different emphases, in accordance with the fields in which he was most active and the paradigm shift dominating those fields. This is overlooked in historical accounts that reduce these theories to a single discipline, whether psychology, sociology, or any other. But other elements, too, played a role in their various approaches.

The fourth question is, namely, to what publics they addressed these theories. In a long-term perspective again, it should be noted that both an academic expert public as well as a broadly educated public had slowly developed during the nineteenth century. We have already said that Sighele, Fournial, and Tarde belonged more or less to the same network. In the relevant chapters we have discussed their connections to the Lombroso and Lacassagne schools, to the publishing houses of Bocca and Alcan, to the Italian and French *Archives of Criminal Anthropology* and related journals, and their participation in (or demonstrative absence from) the successive International Congresses in this field.

Sighele and Fournial wrote their monographs at the completion of their university careers: The books were poorly written and heavily documented; the authors were standing on their toes to impress their academic qualifications on their senior colleagues. Tarde also wrote his first essay on "the crimes of crowds" for a public of experts. But his later essays were better written and more elegantly reasoned, and were clearly meant for a wider educated public: The second appeared in the *Revue des Deux Mondes* and the last two in the *Revue de Paris*. Tarde clearly addressed himself to an elite of opinion leaders in these articles. Le Bon sought an even wider audience. Although he first published his essays on peoples and crowds in the *Revue Scientifique,* his thin books were clearly meant for a much wider public and became true best-sellers, like the more voluminous works of Taine. Whereas the former books were only meant as a contribution to science, then, the latter meant to nourish the public debate.

In this regard, we should conclude with the question What was their respective contribution to crowd psychology, and to social thought in general? As far as the formal object of the field is concerned, Sighele's approach to semiresponsibility has become a regular part of legal thought in many countries. The contributions of Le Bon and Tarde still form the basis for the two major present-day approaches to the field: the deindividuation approach to crowds in a narrower sense, and the emergent norm approach to crowds in a wider sense. Only Fournial was completely forgotten by subsequent generations. As far as the material object of the field is concerned, the successive contributions of Sighele, Le Bon, and Tarde have continually widened the scope of collective behavior studies. In Sighele's monograph the emphasis was still on physically assembled crowds and riots. In Le Bon's monograph the emphasis shifted to social movements and leadership. In Tarde's later essays the emphasis shifted even further, to publics and opinion currents.

They have demonstrated that these seemingly different phenomena often share a common aspect of fluidity: Routine perceptions and behavior give way to contrasting patterns, which may then quickly spread and become adopted on a wider scale. They also emphasized that these phenomena may become more prominent in a mass society of uprooted individuals. Thus both Taine and Le Bon contributed to the theory of totalitarianism.

Finally, we may ask why crowd psychology never took hold as a universally recognized academic field. One reason may be the scale and volatility of the most dramatic manifestations of collective behavior. This makes it difficult to study it in a controlled experimental situation, or even to collect reliable statistical data after the fact – though some attempts have been made at both. Thus crowd psychology has long been superseded by other subjects such as group dynamics and organization research, by opinion polling and attitude scaling.

Another reason may be the imperative of a multidisciplinary approach. With the strict division of labor between psychology and sociology, between various disciplines and subdisciplines continuing unabated, this often remains an unattainable ideal. Funds for a truly multifaceted (though not always integrated) study are usually made available to "task forces," set up after some collective incident such as a riot has taken place. But there is little systematic effort to enhance our understanding of these enigmatic phenomena. In this sense, crowd psychology still has a long way to go.

Bibliography

Unpublished material

The correspondence in private hands has not been numbered, and is usually just kept in alphabetical or chronological order.

Sighele

Interview with his great-niece Anna Maria Gadda Conti Castellini in her Milan apartment, April 1982.

Sighele's lecture notes in the Museo del Risorgimento in Trento:

1ª Lezione: La sociologia e la psicologia collettiva (32 handwritten pages)
2ª Lezione: L'anima della folla (35 handwritten pages)
3ª Lezione: L'intelligenza della folla (38 handwritten pages)
4ª Lezione: La vita pubblica e la psicologia collettiva (37 handwritten pages)
5ª Lezione: Il problema morale dell'anima collettiva (39 handwritten pages); Il problema morale dell'anima collettiva (another version, 44 handwritten pages).

Fournial

Interview with his grand-nephew Charles Fournial in his Draguignan apartment, July 13, 1990.

Personal file. Archives, Military Hospital Val de Grâce, Paris.

Le Bon

Papers in possession of Pierre Duverger, Paris, including:

Letter from Tarde to Le Bon, March 13, 1899
Notebook with successive outlines for the *Psychologie des Foules*.

Tarde

Interview with his son Guillaume de Tarde in his Paris apartment, November 18, 1981.
Correspondence in possession of his granddaughter, Mrs. Paul Bergeret de Tarde, Paris, including:

Letter from Sighele to Tarde, October 17, 1890
Letter from Ribot to Tarde, March 30, 1892
Letter from Le Bon to Tarde, July 7, 1894
Letter from Sighele to Tarde, February 27, 1896
Letter from Sighele to Tarde, March 5, 1896
Letter from Le Bon to Tarde, April 17, 1897
Letter from Le Bon to Tarde, May 1898
Letter from Le Bon to Tarde, June 1898
Letter from Le Bon to Tarde, July 1898
Letter from Le Bon to Tarde, Nov. 26, 1899.

Manuscript of "L'opinion et la conversation" in the Lacassagne Archives, Lyons.

Published material

If two years (separated by a slash) are given for one title, the first year indicates the original edition and the second year, the edition used. If two titles are given, one is the original and the other, the translation used.

Abendroth, W. (1972). *Sociale geschiedenis van de Europese arbeidersbeweging*. Nijmegen: Sun. (Original: *Sozialgeschichte der Europäischen Arbeiterbewegung*. Frankfurt am Main: Suhrkamp, 1965.)
Abraham, J. H. (1973). *Origins and growth of sociology*. Harmondsworth: Penguin.
Accattatis, V. (1979). Introduzione. In E. Ferri, *Sociologia criminale*. Milan: Feltrinelli, pp. 9–51.
Actes du Premier Congrès International d'Anthropologie Criminelle (Roma 1885). (1886–7). Turin: Bocca.
Actes du Deuxième Congrès International d'Anthropologie Criminelle (Paris 1889). (1890). Lyon: Storck.
Actes du Troisième Congrès International d'Anthropologie Criminelle (Bruxelles 1892). (1893). Brussels: Hayer.
Actes du Quatrième Congrès International d'Anthropologie Criminelle (Genève 1896). (1897). Geneva: Kündig.

Actes du Cinquième Congrès International d'Anthropologie Criminelle (Amsterdam 1901). (1901). Amsterdam: De Bussy.

Actes du Premier Congrès International de l'Hypnotisme (Paris 1889). (1890). Paris: Doin.

Agathon (pseud. of Henri Massis and Alfred de Tarde). (1913). *Les jeunes gens d'aujourdhui.* Paris: Plon.

Ahrendt, H. (1951/1971). *The origins of totalitarianism.* New York: Meridian.

Albano, G., et al. (1892). Lettre au président du comité d'organisation du Troisième Congrès International d'Anthropologie Criminelle. *La Scuola Positiva,* 2: 423–4.

Albertoni, E. A. (1973). *Il pensiero politico di Gaetano Mosca – valori, miti, ideologia.* Milan: Cisalpino-Goliardica.

Albrecht-Carrié, R. (1968). *Italy – From Napoleon to Mussolini.* 6th ed. New York: Columbia Univ. Press.

Alff, W. (1973). Die Associazione Nazionalista Italiana von 1910. In *Der Begriff Faschismus und andere Aufsätze zur Zeitgeschichte.* Frankfurt am Main: Suhrkamp.

Allport, G. (1954). The historical background of modern social psychology. In G. Lindzey, *Handbook of social psychology.* Vol. I, Cambridge, Mass.: Addison-Wesley, pp. 3–56.

Ancel, M., et al. (1977). Le centenaire de "L'uomo delinquente" de Lombroso. *Revue de science criminelle et de droit pénal comparé.* New series, No. 1.

Andreucci, F., et al., eds. (1979). *Il movimento operaio italiano – Dizionario biografico 1853–1943.* Rome: Riuniti.

Apfelbaum, E. (1981). Origines de la psychologie sociale – Le cas d'Alfred Binet. In P. Besnard, ed.

Apfelbaum, E., and G. McGuire. (1986). Models of suggestive influence and the disqualification of the crowd. In C. F. Graumann and S. Moscovici, eds., 1986a.

Aron, R. (1967/1970). *Main currents of sociological thought.* 2 vols. Harmondsworth, U.K.: Penguin.

Aubry, P. (1887/1894). *La contagion du meurtre: Étude d'anthropologie criminelle.* Paris: Alcan.

Aulard, A. (1907/1908). *Taine – historien de la Révolution Française.* Paris: Colin.

Aymard, M., and J. Georgelin, P. Guichonnet, P. Racine, F. Thiriet (1977). *Lexique historique de l'Italie.* Paris: Colin.

Azam, E. (1887). *Hypnotisme et double conscience.* Paris: Alcan.

Azéma, J. P., and M. Winock. (1976). *La Troisième République 1870–1940.* Paris: Calmann-Lévy.

Bachrach, P. (1970). *Die Theorie demokratischer Eliten-Herrschaft.* Frankfurt: Europäische Verlags Anstalt.

Bagehot, W. (1869/1873). *Physics and politics – Or thoughts on the application of the principles of 'natural selection' and 'selection' to political society*. London: King.

Baldick, R. (1971/1973). *Dinner at Magny's*. Harmondsworth, U.K.: Penguin.

Baldwin, J. M. (1897). *Social and ethical interpretations in mental development – A study in social psychology*. New York: Macmillan.

Barnes, B. (1974). *Scientific knowledge and sociological theory*. London: Routledge & Kegan Paul.

(1977). *Interests and the growth of knowledge*. London: Routledge & Kegan Paul.

(1982). *T. S. Kuhn and social science*. New York: Columbia Univ. Press.

Barnes, H. E., ed. (1961). *Introduction to the history of sociology*. Chicago: Univ. of Chicago Press.

Barzal, P. (1968). *Les fondateurs de la Troisième République*. Paris: Colin.

Barrows, S. (1981). *Distorting mirrors – Visions of the crowd in late nineteenth century France*. New Haven, Conn.: Yale Univ. Press.

Barrucand, D. (1967). *Histoire de l'hypnose en France*. Paris: Presses Universitaires de France.

Baschwitz, K. (1940/1951). *Denkend mens en menigte – Bijdrage tot een exacte massapsychologie*. The Hague: Leopold. (Original: *Du und die Masse. Studien zu einer exakten Massenpsychologie*. Amsterdam: Feikema & Carelsen, 1938.)

Bataille, A. (1881–99). *Causes criminelles et mondaines de 1880–1897/98*. 20 vols. Paris: Dentu.

Bélugou, L. (1897). Enquête sur l'oeuvre de Taine. *Revue Blanche*, August 15.

Bem, S., H. Rappard, and W. van Hoorn, eds. (1983–7). *Studies in the history of psychology and the behavioral sciences*. (Proceedings of the annual meetings of Cheiron, European Society for the History of the Behavioral and Social Sciences.) Leiden: Psychological Inst./Rijksuniversiteit.

Bem, S. (1985). *Het bewustzijn te lijf – Een geschiedenis van de psychologie*. Amsterdam: Boom.

Bender, D. (1965). The development of French anthropology. *Journal of the History of the Behavioral Sciences*, 1: 139–52.

Bérillon, E. (1891). *Hypnotisme et suggestion*. Paris: Société des Éd. Scientifiques.

Bernheim, H. (1884). *De la suggestion dans l'état hypnotique et dans l'état de veille*. Paris: Doin.

(1886). *De la suggestion et de ses applications à la thérapeutique*. Paris: Doin.

(1891). *Hypnotisme, suggestion, psychothérapie*. Paris: Doin.

Besnard, P., ed. (1981). Sociologies françaises au tournant du siècle – Les concurrents du groupe durkheimien. *Revue française de sociologie*, 22, No. 3.

Bianchi, L., ed., et al. (1908). *L'opera di Cesare Lombroso*. Turin: Bocca.

Binet, A. (1892). *Les altérations de la personnalité*. Paris: Alcan.

Blanchard, M., and E. Giscard d'Estaing. (n.d.). *Gustave Le Bon et le socialisme*. Toulon/St. Roch: Bourély.

Blondel, C. (1928/1946). *Introduction à la psychologie collective*. Paris: Colin.

Blumer, H. (1939/1969). Collective behavior. In A. McClung Lee, ed., *Principles of sociology*. New York: Barnes & Noble.

Bobbio, N. (1972). *On Mosca and Pareto*. Geneva: Droz.

Boef, C. (1984). Van Massapsychologie tot Collectief Gedrag – De Ontwikkeling van een Paradigma. Leiden: diss.

Bogardus, E. S. (1940/1948). *The development of social thought*. New York: Longmans, Green.

Bohn, G., et al. (1934). *La foule*. Paris: Alcan.

Boon, L. (1982). *Geschiedenis van de psychologie*. Amsterdam: Boom.

Bosworth, R. J. B. (1979). *Italy, the least of the great powers: Italian foreign policy before the First World War*. Cambridge: Cambridge Univ. Press.

Bottomore, T. (1964/1973). *Elites and society*. Harmondsworth, U.K.: Penguin.

Boudon, R., F. Bourricaud, and A. Girard, eds. (1981). *Science et théorie de l'opinion publique – Hommage à Jean Stoetzel*. Paris: Retz.

Bourgin, G. (1967). *La Troisième République 1870–1914*. Paris: Colin.

Boutmy, E. (1901/1909). *Essai d'une psychologie politique du peuple anglais au XIXe siècle*. 3rd ed. Paris: Colin.

(1902/1911). *Éléments d'une psychologie politique du peuple américain*. 3rd ed. Paris: Colin.

Bouvier, J., ed. (1964). *Les deux scandales de Panama*. Paris: Julliard.

Bramson, L. (1961). *The political context of sociology*. Princeton, N.J.: Princeton Univ. Press.

Brandell, G. (1967). *Freud – Enfant de son siècle*. Paris: Minard.

Brévan, B. (1979). Relire Gabriel Tarde. *Ethnopsychologie – Revue de la psychologie des peuples*. 34 (No. 2): 239–51.

Brozek, J., and N. Dazzi. (1977). Contemporary historiography of psychology: Italy. *Journal of the History of the Behavioral Sciences*, 13: 33–40.

Bruno, G. (1877/1912). *Le tour de France par deux enfants*. Paris: Bélin.

Bryder, T. (1986). Political psychology in Western Europe. Ch. 15 in M. Hermann, ed., *Political psychology*. San Francisco: Jossey-Bass.

Bulferetti, L. (1975). *Lombroso*. Turin: Utet.

Burnham, J. (1943/1970). *The Machiavellians – Defenders of freedom*. Chicago: Regnery/Gateway.

Busino, G. (1967). *Introduction à une histoire de la sociologie de Pareto.* Geneva: Droz.

Bynum, W. F. & R. Porter, and M. Shepherd. (1985). *The anatomy of madness: Essays in the history of psychiatry.* London: Tavistock.

Castellini, G. (1913). Scipio Sighele. *Rassegna contemporanea.* Anno VI, Serie II, Fasc. XXIII. (Also as a brochure published by Bontempelli, Rome.)

Charcot, J. M. (1886–90). *Oeuvres complètes.* Vols. 1–9. Paris: Delahaye & Lecrosnier.

Charlton, D. G. (1959). *Positivist thought in France during the Second Empire, 1852–1870.* Oxford: Clarendon Press.

Cipolla, C. M., ed., (1973). *The emergence of industrial societies.* Economic history of Europe, 2 vols. London: Collins/Fontana.

Clark, T. [N.] (1969). *Gabriel Tarde on communication and social influence.* Chicago: Univ. of Chicago Press.

(1973). *Prophets and patrons – The French university and the emergence of the social sciences.* Cambridge, Mass.: Harvard Univ. Press.

Cochart, D. (1982). Les foules et la Commune – Analyse des premiers écrits de psychologie des foules. *Recherches de psychologie sociale,* 4: 49–60.

Cooter, R. (1984). *The cultural meaning of popular science: Phrenology and the organization of consent in nineteenth-century Britain.* Cambridge: Cambridge Univ. Press.

Coser, L. (1971). *Masters of sociological thought.* New York: Harcourt Brace Jovanovich.

Croce, B. (1929). *A history of Italy, 1871–1915.* Oxford: Clarendon Press. (Original: *Storia d'Italia dal 1871 al 1915.* Rome: Laterza, 1927–28, 1977.)

Curtis, M. (1959). *Three against the republic: Sorel, Barrès and Maurras.* Princeton, N.J.: Princeton Univ. Press.

Dansette, A. (1946). *Le Boulangisme.* Paris: Fayard.

Darwin, C. (1859/1974). *The origin of species by means of natural selection, or: The preservation of favoured races in the struggle for life.* Harmondsworth, U.K.: Pelican.

(1871/1877). *The descent of man and selection in relation to sex.* 2d. ed. London: Murray.

Davis, M. M. (1909). *Psychological interpretations of society.* New York: Columbia Univ./Longmans, Green.

Debray, R. (1979). *Le pouvoir intellectuel en France.* Paris: Ramsay.

De Grand, A. J. (1978). *The Italian Nationalist Association and the rise of fascism in Italy.* Lincoln: Univ. of Nebraska Press.

Les déjeuners hebdomadaires de Gustave Le Bon. 1928. Paris: Flammarion.

Del Carria, R. (1976). *Proletari senza rivoluzione – Storia delle classi subalterne in Italia.* 2 vols. Rome: Savelli.

De Maupassant, G. (1888/1921). *Sur l'eau*. Paris: Conard. (Engl. transl. *Afloat*. London: Routledge, 1889.)

Den Boer, P. (1987). *Geschiedenis als beroep – De professionalisering van de geschiedbeoefening in Frankrijk 1818–1914*. Nijmegen: Sun.

De Pater, B. (1987). De Tour de France in de schoolbanken. *NRC Handelsblad*, February 19, p. B1.

Des Cilleuls, J. (1956). À propos de la mission Foureau-Lamy et de l'oeuvre des médecins transsahariens. *Médecine Militaire Française*, 2: 50–2.

(1956). À propos de la mission du Congo-Nil: le souvenir du Dr. J. Emily. *Histoire de la Médecine*, 6(No. 6): 50–3.

Despy-Meyer, A. (1973). *Inventaire des archives de l'Université Nouvelle de Bruxelles*. Brussels: Inst. des Hautes Études de Belgique & Université Libre de Bruxelles.

Despy-Meyer, A., and P. Goffin. (1976). *Liber memorialis de l'Institut des Hautes Études de Belgique*. Brussels: Inst. des Hautes Études de Belgigue & Université Libre/Service des Archives.

De Tarde, A., a.o. (1909). *Gabriel Tarde – Introduction et pages choisis par ses fils*. Paris: Michaud.

Diener, E. (1976). Effects of prior destructive behavior, anonymity, and group presence on deindividuation and aggression. *Journal of Personality and Social Psychology*, 33(No. 5): 497–507.

(1979). Deindividuation, self-awareness and disinhibition. *Journal of Personality and Social Psychology*, 37(No. 7): 1160–71.

Digeon, C. (1959). *La crise allemande de la pensée française*. Paris: Presses Universitaires de France.

Dommanget, M. (1938). *De la Marseillaise à l'Internationale*. Paris: Libr. Populaire.

(1967). *Histoire du drapeau rouge*. Paris: Éd./Libr. de l'Étoile.

(1972). *Histoire du Premier Mai*. Paris: Éd. de la Tête des Feuilles.

Droz, J., ed. (1974). *Histoire générale du socialisme*. Vol. 2: 1875–1918. Paris: Presses Universitaires de France.

Droz, N. (1895). *Études et portraits politiques*. Geneva/Paris: Eggiman/Alcan.

Du Camp, M. (1878–80). *Les convulsions de Paris*. 4 vols. Paris: Hachette.

Duchène-Marullaz. Sur le terre marocaine – Une inauguration à l'hôpital militaire de Fez. *La Caducée – Colonies, Guerre, Marine*. October 5, 1912, pp. 1 ff.

Dumas, G., ed. (1923–4). *Traité de psychologie*. 2 vols. Paris: Alcan.

Durkheim, E. (1895/1967). *Les règles de la méthode sociologique*. Paris: Presses Universitaires de France.

(1897/1967). *Le suicide – Étude de sociologie*. Paris: Presses Universitaires de France.

Duverger, P. (1984). *Matière égal énergie – C'est Gustave Le Bon*. Paris: Les Amis de Gustave Le Bon.
Ellenberger, H. (1970). *The discovery of the unconscious – The history and evolution of dynamic psychiatry*. New York: Basic.
Elwitt, S. (1974). *The making of the French Third Republic – Class and politics in France, 1868–1884*. Baton Rouge: Louisiana State Univ. Press.
Espinas, A. (1877/1924). *Des sociétés animales*. 3d ed. Paris: Alcan.
(1910). *Inauguration du monument élevé à la mémoire de G. Tarde à Sarlat*. Séances et travaux de l'Académie des Sciences Morales et Politiques (Institut de France), 174 (N.S. 74).
Esquirol, J. (1838). *Des maladies mentales*. Paris: Baillière.
Essertier, D. (1930). *Philosophes et savants du XXe siècle*. Vol. 5: *La Sociologie*. Paris: Alcan.
Evans, R., ed. (1969). *Readings in collective behavior*. Chicago: Rand McNally.
Favre, P. (1981). Émile Boutmy et l'École Libre des Sciences Politiques. In Besnard, ed.
(1983). Gabriel Tarde et la mauvaise fortune d'un "baptême sociologique" de la science politique. *Revue Française de Sociologie*, 24: 3–30.
Ferri, E. (1884). *I nuovi orizzonti del diritto e della procedura penale*. Bologna: Zanichelli. Later: *Sociologia criminale* (English transl.: *Criminal sociology*. London: Fisher Unwin, 1895, abr. ed.).
(1894). *Socialismo e scienza positiva – Darwin, Spencer, Marx*. Rome: Casa Editrice Italiana.
et al. (1895). Polemica sulla "Psychologie des foules." *La Scuola Positiva*, 5: 367–75.
(1899). *Difese penali e studi di giurisprudenza* Turin: Fratelli Bocca.
(1903). Autobiografia. Suppl. alla *Folla*, No. 43, pp. 2–16.
Fest, J. C. (n.d.). *Hitler – Een biografie*. Baarn: In den Toren. (Original: *Hitler – Eine Biographie*. Frankfurt am Main: Ullstein.)
Festinger, L., A. Pepitone, and T. Newcomb. (1952). Some consequences of deindividuation in a group. *Journal of Abnormal and Social Psychology*, 47: 382–9.
Finer, S., ed. (1966/1976). *Vilfredo Pareto – Sociological writings*. Oxford: Blackwell.
Fischer, R. (1961). *Masse und Vermassung*. Zürich: Polygraphischer Verlag.
Fouillée, A. (1890). *L'évolutionnisme des idées forces*. Paris: Alcan.
(1892–93). *La psychologie des idées forces*. Paris: Alcan.
Fournial, H. (1892). *Essai sur la psychologie des foules – Considérations médico-judiciaires sur les responsabilités collectives*. Lyon: Storck; Paris: Masson.

Franchi, B. (1908). *Enrico Ferri – Il noto, il mal noto e l'ignorato*. Turin: Bocca.
Frary, R. (1981). *Du bon usage de la mauvaise foi*. Paris: Aubier Montaigne. (Original: *Manuel du Démagogue*. Paris: Libr. Cerf, 1884.)
Frémy, D., and M. Frémy. (1977). *Quid 1978*. Paris: Laffont.
Freud, S. (1953–74). *Complete psychological works – The standard edition*. 24 vols. London: Hogarth Press & The Inst. of Psycho-analysis.
Freund, J. (1974). *Pareto – La théorie de l'équilibre*. Paris: Séghers.
Gagnon, P. A. (1964). *France since 1789*. New York: Harper & Row.
Garbari, M. (1974). Il pensiero politico di Scipio Sighele. *Rassegna Storica del Risorgimento*. 61(No. 3): 391–426, and (No. 4): 523–61.
(1977). *L'età giolittiana nelle lettere di Scipio Sighele*. Trento: Società di Studi Trentini di Scienze Storiche.
Garrett, M. B., and J. L. Godfrey (1947). *Europe since 1815*. New York: Appleton Century Crofts.
Gauthier, R., ed. (1965). *"Dreyfusards" – Souvenirs de Mathieu Dreyfus et autres inédits*. Paris: Julliard.
Geiger, R. L. (1975). The institutionalization of sociological paradigms: Three examples from early French sociology. *Journal of the History of the Behavioral Sciences*, 2: 235–45.
(1977). Democracy and the crowd: The social history of an idea in France and Italy, 1890–1914. *Societas – A Review of Social History*, 7(No. 1): 47–71.
(1981). René Worms et l'organisation de la sociologie. In: Besnard, ed.
Gérard, A. (1970). *La Révolution Française: mythes et interprétations 1789–1970*. Paris: Flammarion.
Ghisalberti, C. (1974). *Storia costituzionale d'Italia 1849–1948*. Rome: Laterza.
Giachetti, C. (1914). Scipio Sighele. *Nuova Antologia*, 49 (No. 1009; Jan. 1): 94–102.
Giner, S. (1976). *Mass society*. London: M. Robertson.
Girardet, R., ed. (1966). *Le nationalisme français 1871–1914, Textes choisis*. Paris: Armand Colin.
Giraud, V. (1902). *Essai sur Taine, son oeuvre et son influence d'après des documents inédits*. Paris: Hachette.
Gisevius, H. B. (1973). *Adolf Hitler*. De Bilt: De Fontein. (Original: *Adolf Hitler*. Munich: Rütten & Loening, 1963.)
Goddijn, H. P. M. (1973). *Sociologie, socialisme en democratie – De politieke sociologie van Émile Durkheim*. Meppel: Boom.
Godechot, J. (1974). *Un jury pour la révolution*. Paris: Laffont.
Goffin, P. (1969). *Histoire de l'Institut des Hautes Études de Belgique*. Brussels: IHEB.

Gould, S. (1981/1984). *The mismeasure of man*. Harmondsworth, U.K.: Penguin/Pelican.

Graumann, C. F., & S. Moscovici, eds. (1986a). *Changing conceptions of crowd mind and behavior*. New York: Springer.

(1986b), *Changing conceptions of leadership*. New York: Springer.

Gregor, A. J. (1969). *The ideology of fascism – The rationale of totalitarianism*. New York: Free Press.

(1979). *Young Mussolini and the intellectual origins of fascism*. Berkeley: Univ. of California Press.

Grunenberg, A., ed. (1970). *Die Massenstreikdebatte*. Frankfurt am Main: Europäische Verlags Anstalt.

Grupp, S. E., ed. (1968). *The positive school of criminology – Three lectures by Enrico Ferri*. Pittsburgh: Pittsburgh Univ. Press.

Grupp, S. E. (1968). The sociology of Gabriel Tarde. *Sociology and Social Research*, 52 (No. 4): 333–47.

Guichonnet, P. (1969). *La monarchie libérale 1870–1922*. (*Histoire d'Italie*, Vol. 1). Paris: Hatier.

Guillain, G. (1955). *J. M. Charcot – Sa vie, son oeuvre*. Paris: Masson.

Gurvitch, G., ed. (1963). *Traité de sociologie*. Paris: Presses Universitaires de France.

Halasz, N. (1955). *Captain Dreyfus – The story of a mass hysteria*. New York: Simon & Schuster.

Hammond, M. (1980). Anthropology as a weapon of social combat in late 19th century France. *Journal of the History of the Behavioral Sciences*, 16: 118–32.

Hecker, J. F. C. (1832). *Die Tanzwuth – Eine Volkskrankheit im Mittelalter*. Berlin: T. C. F. Enslin.

Heilbron, J. (1983). *Sociologie in Frankrijk – Een historische schets*. Amsterdam: Siswo (publ. 270).

Hillman, R. G. (1965). A scientific study of mystery: the role of the medical and popular press in the Nancy–Salpêtrière controversy on hypnotism. *Bulletin of the History of Medicine*, 39: 162–82.

Hobsbawm, E. J. (1959/1971). *Primitive rebels – Studies in archaic forms of social movement in the 19th and 20th centuries*. Manchester: Manchester Univ. Press.

Hobsbawm, E. (1969/1973). *Bandieten*. Amsterdam: Wetenschappelijke Uitgeverij.

Hommage à Fernand Foureau – Mémento de la mission saharienne 1898–1900. (1914). Bourges: Foucrier.

Hughes, H. S. (1958/1974). *Consciousness and society – The reorientation of European social thought 1890–1930*. New York: Vintage.

Jaeger, S., and I. Staeble. *Die gesellschaftliche Genese der psychologie*. Frankfurt: Campus.

Janet, P. (1889/1894). *L'automatisme psychologique*. 2nd ed. Paris: Alcan.
Jonas, F. (1972–4). *Geschichte der Soziologie*. 4 vols. Reinbek/Hamburg: Rowohlt.
Keller, W. (1966). *Und wurden zerstreut unter alle Völker – Die nachbiblische Geschichte des jüdischen Volkes*. Munich: Knaur; Zürich: Droemer.
Kinder, H., and W. Hilgemann, eds. (1966). *Atlas zur Weltgeschichte*. Munich: Deutscher Taschenbuch Verlag.
Knorr-Cetina, K. D. (1981). *The manufacture of knowledge – An essay on the constructivist and contextual nature of science*. Oxford: Pergamon.
Knutson, J., ed. (1973). *Handbook of political psychology*. San Fransisco: Jossey-Bass.
Kuczynski, J. (1967). *Die Geschichte der Lage der Arbeiter unter dem Kapitalismus*. 38 vols. Berlin: Akademie.
Kuhn, T. S. (1962). *The structure of scientific revolutions*. Chicago: Univ. of Chicago Press.
Kühnl, R. (1971). *Formen bürgerlicher Herrschaft: Liberalismus – Faschismus*. Reinbek/Hamburg: Rowohlt.
Laborde, J. B. V. (1872). *Les hommes et les actes de l'insurrection de Paris devant la psychologie morbide – Lettres à M. le Docteur Moreau*. Paris: Germer Baillière.
Lacassagne, A. (1882). *L'homme criminel comparé à l'homme primitif*. Lyon: Société d'Anthropologie/Association Typographique.
(1890). Préface. In E. Laurent, *Les habitués des prisons de Paris*. Lyon: Storck; Paris: Masson.
Lacassagne, A., et al. (1894). *L'assassinat du président Carnot*. Lyon: Storck; Paris: Masson.
Lacassagne, A. (1904). Gabriel Tarde. *Archives d'Anthropologie Criminelle*, 19(Nos. 127–8): 501–34.
Landolfi, E. (1981). *Scipio Sighele – Un giobertiano tra democrazia nazionale e socialismo tricolore*. Rome: Volpi.
Lang, K., and G. E. Lang. (1961). *Collective dynamics*. New York: Crowell.
Larousse de la langue française/Lexis. (1977). Paris: Libr. Larousse.
Latour, B. (1987). *Science in action*. Milton Keynes, U.S.: Open Univ. Press.
Latour, B., and S. Woolgar. (1979/1986). *Laboratory life – The construction of scientific facts*. Princeton, N.J.: Princeton University Press.
Lazzeroni, V. (1972). *Le origini della psicologia contemporanea*. Florence: Giunti Barbèra.
Le Bon, G. (1872). *La vie – Physiologie humaine*. Paris: Rothschild.
(1881). *L'homme et les sociétés*. Paris: Rothschild.
(1888/1985). *Rôle des juifs dans la civilisation*. Paris: Les Amis de Gustave

Le Bon. (Original in *Revue Scientifique*, September 29, 1888. Later in: *Les premières civilisations*.)

(1889). *Les premières civilizations*. Paris: Marpon & Flammarion.

(1894a). La lutte guerrière des peuples & La lutte économique des races. *Revue Scientifique*, 31 (4e série, Tome I; No. 7, Feb. 17): 193–200.

(1894b). *Lois psychologiques de l'évolution des peuples*. Paris: Alcan. (Engl. transl.: *The psychology of peoples*. London: Fisher Unwin, 1899.)

(1895a/1963). *Psychologie des Foules*. Paris: Presses Universitaires de France (originally: Alcan). (English transl.: *The crowd – A study of the popular mind*. New York: Viking/Compass, 1966.)

(1895b). À propos des foules criminelles. *Revue Scientifique*, 32: 635.

(1898/1977). *Psychologie du socialisme*. Paris: Les Amis de Gustave Le Bon (originally: Alcan). (English transl. *Psychology of socialism*. London: Fisher Unwin, 1899.)

(1902). *Psychologie de l'éducation*. Paris: Flammarion.

(1910/1926). *La psychologie politique*. Paris: Flammarion.

(1911/1917). *Les opinions et les croyances*. Paris: Flammarion.

(1912). *La Révolution Française et la psychologie des révolutions*. Paris: Flammarion. (Engl. transl. *The French Revolution and the psychology of revolution*. Republ. New Brunswick/London: Transaction Books, 1980).

(1913-24/1978). *Aphorismes du temps présent & Les incertitudes de l'heure présente* (part.). Paris: Les Amis de Gustave Le Bon (with *Lois psychologiques de l'évolution des peuples*).

(1931). *Bases scientifiques d'une philosophie de l'histoire*. Paris: Flammarion.

Lefebvre, G. (1954). *Études sur la Révolution Française*. Paris: Presses Universitaires de France.

Lenin, V. I. (1961–81). *Collected works*. Moscow: Progress.

Lentini, O. (1971/71). Storiografia della sociologia italiana, 1860–1925. *Critica Sociologica*, 20: 116–40.

Léon, A. (1942/1968). *La conception matérialiste de la question juive*. Paris: Études et Documentation Internationales.

Letourneau, C. (1878). *Physiologie des passions*. 2d ed. Paris: Reinwald.

Loewenberg, P. (1983). *Decoding the past – The psychohistorical approach*. New York: Knopf.

Lombroso, C. (1876). *L'uomo delinquente – studiato in rapporto alla antropologia, alla medicina legale ed alle discipline carcerarie*. Milan: Hoepli.

(1894). *Gli anarchici*. Turin: Bocca.

(1899). *Le crime – Causes et remèdes*. Paris: Reinwald/Schleicher.

Lombroso, C. and R. Laschi. (1890). *Il delitto politico e le rivoluzioni in rapporto al diritto, all'antropologia criminale ed alla scienza di governo*. Turin: Bocca.

Long, S., ed. (1981). *The handbook of political behavior*. New York/London: Plenum.
Lorenz, C. (1987). *De constructie van het verleden*. Amsterdam: Boom.
Lubek, I. (1981). La psychologie sociale perdue de Gabriel Tarde. In Besnard, ed.
Lück, H., R. Miller, and W. Rechtien, eds. (1984). *Geschichte der Psychologie*. Munich: Urban & Schwarzenberg.
Lück, H., et al. (1987). *Sozialgeschichte der Psychologie*. Opladen: Leske & Budrich.
Lukes, S. (1973/1981). *Émile Durkheim – His life and work*. Harmondsworth, U.K.: Penguin.
Luys, J. (1894). Études de psychologie sociale – La foule criminelle. *Annales de Psychiatrie et d'Hypnologie*, 4: 289–97.
McClelland, J. S. (1971). *The French right – From De Maistre to Maurras*. London: Cape.
(1989). *The crowd and the mob – From Plato to Canetti*. London: Unwin Hyman.
McDougall, W. (1908, 1914). *An introduction to social psychology*. Boston: Luce.
McGuire, G. (1984). The collective unconscious – Psychical research in French psychology, 1880–1920. Paper presented at a symposium on Controversies in psychology during France's Belle Époque, Toronto.
(1985). Gustave Le Bon and the irrationalism of the social crowd. Unpublished manuscript, York Univ., Toronto.
(1986a). Pathological subconscious and irrational determinism in the social psychology of the crowd – The legacy of Gustave Le Bon. In W. J. Baker, M. V. Hyland, H. van Rappard, and A. W. Staats, eds., *Current issues in theoretical psychology*. Amsterdam: North Holland.
(1986b). Review of: Serge Moscovici – *The age of the crowd* (unpubished).
(1986c). Psychopathology, hypnosis and Pierre Janet's conception of the organic subconscious. Paper presented at the 18th annual meeting of Cheiron, Univ. of Guelph.
(1986d). Hypnotic suggestion and criminal responsibility: Free will and individual agency in late 19th century French psychopathology. In S. Bem and H. Rappard (1986).
Mackay, C. (1841, 1980). *Extraordinary delusions and the madness of crowds*. London: Bentley; New York: Crown.
Maisonneuve, J. (1960). *Psychologie sociale*. Paris: Presses Universitaires de France.
Maitron, J., ed. (1964). *Ravachol et les anarchistes*. Paris: Julliard.
Malcangi, G. (1972). Giuseppe Alberto Pugliese e il casato Nencha. *Il Tranesiere* (Periodico Quindicinale), 29(No. 10): 5–9.

Mannheim, H. (1960). *Pioneers in criminology*. London: Stevens.
(1966/1974). *Vergleichende Kriminologie*. Stuttgart: Enke. Original: *Comparative criminology*. 2nd ed. London: Routledge & Kegan Paul.
(1975). Criminology. In *Encyclopaedia Britannica*, Vol. 5, Macropedia, pp. 282–5.
Marhaba, S. (1981). *Lineamenti della psicologia italiana*, 1870–1945. Florence: Giunti Barbèra.
Marx, K., and Engels, F. (1974–). *Collected works*. London: Lawrence & Wishart.
Marzani, G. (1953). Scipio Sighele. *Studi Trentini di Scienze Storiche*, 32: 463–74.
Maser, W. (1973). *Hitler – Legende, mythe, werkelijkheid*. Amsterdam: Arbeiderspers. Original: *Adolf Hitler – Legende, Mythos, Wirklichkeit*. Munich/Esslingen: Bechtle, 1971.
Matagrin, A. (1910). *La psychologie sociale de Gabriel Tarde*. Paris: Alcan.
Maury, A. (1861/1878). *Le sommeil et les rêves – Études psychologiques*. 4th ed. Paris: Didier.
Mayeur, J. M. (1973). *Les débuts de la IIIe République*. Paris: Seuil.
Mead, G. H. (1956/1977). *On social psychology*. Chicago: Univ. of Chicago Press.
Meisel, J. H. (1956). *The myth of the ruling class – Gaetano Mosca and the "Elite."* Ann Arbor: Univ. of Michigan Press.
(1965). *Pareto and Mosca*. Englewood Cliffs, N.J.: Prentice-Hall.
Métraux, A. (1983). French crowd psychology: between theory and ideology. In W. Woodward and M. Ash, eds., *The problematic science: Psychology in 19th century thought*. New York: Praeger.
Michels, R. (1908/1921). *Le prolétariat et la bourgeoisie dans le mouvement socialiste italien*. Paris: Giard.
(1911/1968). *Political parties – A sociological study of the oligarchical tendencies of modern democracy*. New York: Macmillan/Free Press. (Original: *Zur Soziologie des Parteiwesens in der modernen Demokratie*. Stuttgart: Kröner.)
(1929). Gaetano Mosca und seine Staatstheorien. *Schmollers Jahrbuch*, 53: 111–30.
Milet, J. (1970). *Gabriel Tarde et la philosophie de l'histoire*. Paris: Vrin.
(1975). Gabriel Tarde et la notion de société. *Recherche Sociale*, 53: 54–69.
Miroglio, A. (1965). *La psychologie des peuples*. Paris: Presses Universitaires de France.
Misiak, H., and V. M. Staudt. (1953). Psychology in Italy. *Psychological Bulletin*, 50 (No. 5): 347–61.
Mitzman, A. (1973). *Sociology and estrangement – Three sociologists in imperial Germany*. New York: Knopf.

Moede, W. (1938). Einführung. In G. Le Bon, *Psychologie der Massen*, pp. i–x. Stuttgart: Kröner.

Mongardini, C. (1965). *Storia e sociologia nell'opera di H. Taine*. Milan: Giuffré.

Morawski, J., ed. (1988). *The rise of experimentation in American psychology*. New Haven, Conn.: Yale Univ. Press.

Mosca, G. (1884). *Sulla teorica dei governi e sul governo parlamentare – Studii storici e sociali*. Palermo: Tip. dello "Statuto."

(1895–96). *Elementi di scienza political*. Rome: Bocca. (English transl.: *The ruling class*. New York: McGraw-Hill, 1939.)

(1933, 1972). *A short history of political philosophy*. New York: Crowell.

Moscovici, S. (1976). *Social influence and social change*. London: Academic Press. (French ed.: *Psychologie des minorités actives*. Paris: Presses Universitaires de France, 1979.)

(1985). *The age of the crowd – A historical treatise on mass psychology*. Cambridge: Cambridge Univ. Press (French ed.: *L'âge des foules – Un traité historique de psychologie des masses*. Paris: Fayard, 1981.)

Moser, H., ed. (1979). *Politische Psychologie*. Weinheim/Basel: Beltz.

Motono, Le Baron. (1914). *L'oeuvre de Gustave Le Bon*. Paris: Flammarion.

Mucchi Faina, A. (1983). *L'abbraccio della folla – Cento anni di psicologia collettiva*. Bologna: Mulino.

Murchison, C., ed. (1930 ff.). *A history of psychology in autobiography*. Worcester, Mass.: Clark Univ. Press.

Naquet, A. (1886). Le parlementarisme. *Revue Bleue*, 23 (No. 25): 769–74; (No. 26): 801–7.

(1887). Le régime représentatif. *Revue Bleue*, 24 (No. 4): 97–103; (No. 5): 138–43.

Néré, J. (1964). *Le Boulangisme et la presse*. Paris: Armand Colin.

Niceforo, A. (1901). Italiani del Nord e Italiani del Sud. Turin: Bocca.

(1910). Antropologia delle classi povere. Milan: Vallardi.

Nietzsche, F. (1886/1968). *Jenseits von Gut und Böse*. Frankfurt: Fischer (Studienausgabe, Vol. 3).

Nijhoff, P. (1980–81). Sorel en de sociologie. *Sociologisch Tijdschrift*, 7: 936–74; 8: 18–47.

Nisbet, R. A. (1966/1973). *The sociological tradition*. London: Heinemann.

Noelle-Neumann, E. (1984). *The spiral of silence – Public opinion, our social skin*. Chicago: Univ. of Chicago Press.

Nolte, E. (1970). *L'action française*. Vol. I. *Le fascisme dans son époque*. Paris: Julliard.

Nye, R. A. (1973). Two paths to a psychology of social action: Gustave Le Bon and Georges Sorel. *Journal of Modern History*, 45 (No. 3): 411–38.

(1975). *The origins of crowd psychology: Gustave Le Bon and the crisis of mass democracy in the Third Republic*. London: Sage.

(1977). *The antidemocratic sources of elite theory: Pareto, Mosca and Michels.* London: Sage.
(1980). Introduction to *The French Revolution. . . .* English transl. of Le Bon, 1912, *La Révolution Française. . . .*
(1983). Review of Barrows (1981) and Moscovici (1981). *Isis,* 74(No. 4): 568–73.
(1984). *Crime, madness, and politics in modern France – The medical concept of national decline.* Princeton, N.J.: Princeton Univ. Press.
Orano, P. (1902). *Psicologia sociale.* Bari: Laterza.
Osborne, T. O. (1974). *The recruitment of the administrative elite in the Third French Republic, 1870–1905. The system of the École Libre des Sciences Politiques.* Ann Arbor, Mich.: Xerox/Univ. Microfilm (Univ. of Connecticut).
Ostrogorski, M. (1903/1979). *La démocratie et les partis politiques.* Paris: Seuil. (English transl.: *Democracy and the organization of political parties,* 1912.)
Padovani, G. (1946). *La stampa periodica italiana di neuropsichiatria e scienze affini.* Milan: Hoepli.
Pannunzio, C. (1945). Italian sociology. Ch. 22 in G. Gurvitch and W. E. Moore, *Twentieth century sociology.* New York: The Philosophical Library, pp. 638–52.
Paoli, B. (1884). *Esposizione storica e scientifica dei lavori di preparazione del Codice Penale Italiano, dal 1866 al 1884.* 2 vols. Florence: Nicolai.
Pareto, V. (1902–3/1965). *Oeuvres complètes,* Vol. V. *Les systèmes socialistes.* Geneva: Droz.
(1962). *Lettre a Maffeo Pantaleoni.* Vol I, 1890–96. Rome: Ed. di Storia e Letteratura.
(1968). *The rise and fall of the elites – An application of theoretical sociology.* Totowa, N.J.: Bedminster Press.
Paris, R. (1968). *Les origines du fascisme.* Paris: Flammarion.
Park, R. (1972). *The crowd and the public.* Chicago: Univ. of Chicago Press. (Original: *Masse und Publikum,* 1904.)
Park, R. E., and E. W. Burgess (1921/1970). *Introduction to the science of sociology.* Chicago: Univ. of Chicago Press.
Pedrotti, G. (1932). Una famiglia di patrioti trentini. *La Lombardia nel Risorgimento Italiano,* 17: 3–66.
Penal code of the Kingdom of Italy. (1931). British Foreign Office translation of the Italian *Codice penale.* London: H.M. Stationery Office.
Perrot, M. (1974). *Les ouvriers en grève – France 1871–1890.* 2 vols. Paris: Mouton.
Picard, E. (1909). *Gustave Le Bon et son oeuvre.* Paris: Mercure de France.
Picavet, F. (1903–04). Notice biographique sur J. G. de Tarde. *Annuaire du Collège de France,* 4: 33–40.

Pinatel, J. (1961). De Lacassagne à la Nouvelle École de Lyon. *Revue de Science Criminelle et de Droit Pénal Comparé,* n.s., No. 1: 151–8.

Pisani-Ferry, F. (1969). *Le général Boulanger.* Paris: Flammarion.

Poliakov, L. (1979). *De arische mythe.* Amsterdam: Arbeiderspers. (Original: *Le mythe arien.* Paris: Calmann-Lévy.)

Porter, T. M. (1986). *The rise of statistical thinking 1820–1900.* Princeton, N.J.: Princeton Univ. Press.

Prins, A. (1884). *Le démocratie et le régime parlementaire.* Brussels: Muquardt.

Procacci, G. (1968). *History of the Italian people.* Harmondsworth, U.K.: Penguin.

Puccini, S., and M. Squillaciotti. (1979). Per un prima ricostruzione critico-bibliografica degli studi demo- etno- antropologici italiani. *Problemi del socialismo,* Series 4a, 20(No. 16): 67–94.

Pugliese, G. (1887). Del delitto collettivo. *Rivista di Giurisprudenza* 12 (Nos. 3–4): 203–26. (Also published as a brochure: Trani, Tip. Valdemaro Vecchi, 1887, 13 pp.)

Rademaker, L. and E. Petersma, eds. (1974). *Hoofdfiguren uit de sociologie.* 2 vols. Utrecht: Spectrum; Antwerp: Aula.

Ranieri, S. (1969). Les développements de l'école positive en Italie. *Revue Internationale de Criminologie et de Police Technique,* 23 (No. 1): 177–84.

Règnard, P. (1887). *Les maladies épidémiques de l'esprit.* Paris: Plon, 1887.

Reibell, E. (1929). *Carnet de route de la mission saharienne Foureau-Lamy, 1898–1900.* Paris: Plon.

(1931). *L'épopée saharienne.* Paris: Plon.

(1935). *Le calvaire de Madagascar – Notes et souvenirs de 1895.* Paris: Berger-Levrault.

Reinach, J. (1890). *La politique opportuniste.* Paris: Charpentier.

(1901–8). *Histoire de l'affaire Dreyfus.* 7 vols. Paris: Éd. de la Revue Blanche.

Reiwald, P. (1952). *De geest der massa.* Bussum: Kroonder. (Original: *Vom Geist der Massen – Handbuch der Massenpsychologie.* Zürich: Pan Verlag, 1946.)

Reuchlin, M. (1957/1980). *Histoire de la psychologie.* 11th ed. Paris: Presses Universitaires de France.

Rhodes, H. T. F. (1956). *Alphonse Bertillon – Father of scientific detection.* London: Harrap.

Ribot, T. (1871). *La psychologie anglaise contemporaine.* Paris: Germer Baillière.

(1873). *L'hérédité – Étude psychologique.* Paris: Ladrange.

(1879). *La psychologie allemande contemporaine.* Paris: Germer Baillière.

(1885). *Les maladies de la personnalité.* Paris: Alcan.

(1889). *Psychologie de l'attention.* Paris: Alcan.
(1896/1925). *La psychologie des sentiments.* 12th ed. Paris: Alcan.
Richet, C. (1884). *L'homme et l'intelligence.* Paris: Alcan.
Ripepe, E. (1971). *Le origini della teoria della classe politica.* Milan: Giuffrè.
Robiquet, P., ed. (1898). *Discours et opinions de Jules Ferry,* Vol. 7. Paris: Colin.
Rodinson, M., et al. (1971). Sur la conception matérialiste de la question juive. *Israc,* No. 5.
Rogger, H., and E. Weber, eds. (1966). *The European Right – A historical profile.* Berkeley: Univ. of California Press.
Roloff, E. A. (1976). *Psychologie der Politik.* Stuttgart: Metzler.
Romano, S. (1977). *Histoire de l'Italie du Risorgimento à nos jours.* Paris: Seuil.
Rosenbauer, M. (1969). *L'École Libre des Sciences Politiques de 1871 à 1896 – L'enseignement des sciences politiques sous la 3e République.* Marburg/Lahn: Philipps Univ. (Philos. Fak., Diss.).
Ross, E. (1908/1925). *Social psychology – An outline and source book.* New York: Macmillan.
Rossi, P. (1898). *L'animo della folla.* Cosenza: Riccio.
(1899a). *Psicologia collettiva – Studii e ricerche.* Cosenza: Riccio.
(1899b). *Mistici e settari – Studio di psico-patologia collettiva.* Cosenza: Riccio.
(1899c). *Giuseppe Mazzini e la scienza moderna – Studio di sociologia e di psicologia.* Cosenza: Caputi.
(1900). Due parole di programma. *Archivio di Psicologia e Scienze Affini,* 1 (No. 1): i–iv.
(1901). *Psicologia collettiva morbosa.* Turin: Bocca.
(1902). *I suggestionari e la folla.* Turin: Bocca.
(1904). *Sociologia e psicologia collettiva.* Rome: Colombo.
Rougerie, J. (1964). *Procès des Communards.* Paris: Julliard.
Rouvier, C. (1986). *Les idées politiques de Gustave Le Bon.* Paris: Presses Universitaires de France.
Rudé, G. (1959/1972). *The crowd in the French Revolution.* London: Oxford Univ. Press.
(1964). *The crowd in history – A study of popular disturbances in France and England, 1730–1848.* New York: Wiley.
(1980). *Ideology and popular protest.* London: Lawrence & Wishart.
Sabatini, G., et al. (1941). *Enrico Ferri – Maestro della scienza criminologica.* Milan: Bocca.
Sahakian, W. S. (1975). *History and systems of psychology.* New York: Schenk/Halsted/Wiley.
(1982). *History and systems of social psychology.* 2nd ed. Washington, D.C.: Hemisphere/McGraw-Hill.

Salomone, W. A. (1960). *Italy in the giollittian era – Italian democracy in the making, 1900–1914*. Philadelphia: Univ. of Pennsylvania Press.

Salomone, W. A., ed. (1970). *Italy from the Risorgimento to fascism*. Garden City, N.Y.: Anchor/Doubleday.

Schérer, E. (1883). *La démocratie en France*. Paris: Libr. Nouvelle. (Engl. transl. *Democracy and France*. London: Richardson, 1884.)

Schumpeter, J. A. (1943, 1974). *Capitalism, socialism and democracy*. London: Unwin.

Sereno, R. (1968). *The rulers – the theory of the ruling class*. New York: Harper & Row.

Sergi, G. (1889). Psicosi epidemica. *Rivista di Filosofia Scientifica*, Series 2a, 8: 151–72. (Also published as a brochure by Dumolard, Milano.)

Sergi, S. (1936/7). Giuseppe Sergi. *Annuario dell'Università di Roma*, pp. 593–8.

Seton-Watson, C. (1967). *Italy from liberalism to fascism, 1870–1925*. London: Methuen.

Shapin, S., and Schaffer, S. (1985). *Leviathan and the air-pump – Hobbes, Boyle, and experimental life*. Princeton, N.J.: Princeton Univ. Press.

Sighele, S. (1891). Il delitto politico. *Archivio Giuridico*, 46, No. 6. (Also published as a brochure by Tip. Fava e Garagnani in Bologna.)

(1891). La folla delinquente. *Archivio di Psichiatria*, 12: 10–53, 222–67, 322–46. (Also published as a book by Frat. Bocca, Turin.) (Fr. ed.: *La foule criminelle*. Paris: Alcan, 1892.)

(1892). *La coppia criminale*. Turin: Bocca. (French ed.: *Le crime à deux*. Paris: Maloine, 1893.)

(1894). *La teorica positiva della complicità*. 2d ed. Turin: Bocca.

(1895). Una pirateria letteraria – A proposito della "Psychologie des Foules" di G. Le Bon. *La Scuola Positiva*, 5: 171–3.

(1897). *La delinquenza settaria*. Milan: Treves. (Fr. ed.: *Psychologie des sectes*. Paris: Giard & Brière, 1898.)

(1901). Le crime collectif. In *Actes du Cinquième Congrès International d'Anthropologie Criminelle*.

(1902). *I delitti della folla*. Turin: Bocca.

(1903). *L'intelligenza della folla*. Turin: Bocca.

(1909). Cesare Lombroso. *Nuova Antologia*, 16 Nov. (Also published as a brochure by Nuova Antologia, Rome.)

(1914). *Letteratura e sociologia – Saggi postumi*. Milan: Treves.

Snyder, L. (1966). *Rassen – Werkelijkheid en waan*. Hilversum: De Haan. (Original: *The idea of racialism*. Princeton, N.J.: Van Nostrand.)

Sorel, G. (1899). Review of G. Le Bon "La psychologie du socialisme." *Revue Internationale de Sociologie*, pp. 152–5.

(1908/1972). *Réflexions sur la violence*. Paris: Éd. Marcel Rivière.

Souchon, H. (1974). Alexander Lacassagne et l'École de Lyon. *Revue de Science Criminelle et de Droit Pénal Comparé*, n.s., No. 3: 533–55.

Squillace, F. (1902). *Le dottrine sociologiche*. Milan: Sandròn.
(1909). *Il problema della psicologia collettiva e sociale, e l'opera di Pasquale Rossi*. Milan/Palermo: Sandròn.
Stein, A. (1955). Adolf Hitler und Gustave Le Bon. *Geschichte in Wissenschaft und Unterricht*, 6: 362–8.
Sternhell, Z. (1978). *La droite révolutionnaire 1885–1914–Les origines françaises du fascisme*. Paris: Seuil.
Stoetzel, J. (1943). *Esquisse d'une théorie des opinions*. Paris: Presses Universitaires de France.
Stone, W. F. (1974). *The psychology of politics*. New York: Free Press.
Streiffeler, F. (1975). *Politische Psychologie – Geschichte und Themen der Theorie des politischen Verhaltens*. Hamburg: Hoffmann and Campe.
Sulloway, F. (1979). *Freud – Biologist of the mind*. New York: Basic.
Szacki, J. (1979). *History of sociological thought*. London: Aldwich.
Taine, H. (1863). Introduction, *Histoire de la littérature anglaise*, Vol. 1. Paris: Hachette, pp. iii–xlviii. (English transl.: *History of English literature*. 2 vols. Edinburgh: Edmonston & Douglas, 1871.)
(1870). *De l'intelligence*. Paris: Hachette. (English transl.: *On intelligence*. London: Reeve, 1871.)
(1872). *Du suffrage universel et de la manière de voter*. Paris: Hachette.
(1875–93). *Les origines de la France contemporaine*. Paris: Hachette. (Abr. English transl.: *The origins of contemporary France*. Chicago: Univ. of Chicago Press, 1974.)
(1894). *Derniers essays de critique et d'histoire*. Paris: Hachette.
(1902–7). *Sa vie et sa correspondance*. 4 vols. Paris: Hachette. (Engl. transl.: *Life and letters of H. Taine*. 3 vols. (slightly abr.). Westminster: Constable, 1902 ff.).
Tanturri, R. (1973). *Pensiero e significato de Gaetano Mosca ed altri scritti*. Padua: Cedam.
Tarde, G. (1889). Review of J. Delboeuf "Le magnétisme animal." *Archives d'Anthropologie Criminelle*, 4: 501–5.
(1890a). *La philosophie pénale*. Lyon: Storck; Paris: Masson.
(1890b/1979). *Les lois de l'imitation*. Paris: Ressources; Genève: Slatkine.
(1890c). Le délit politique. *Revue Philosophique*, 30: 337 ff.
(1891). Études criminelles et pénales. *Revue Philosophique*, 32: 483–506.
(1892). Les crimes des foules. In *Actes du Troisième Congrès d'Anthropologie Criminelle*, pp. 73–90.
(1893). Foules et sectes au point de vue criminel. *Revue des Deux Mondes*, Nov. 15, pp. 349–87.
(1895). *Essais et Mélanges Sociologiques*. Lyon: Storck; Paris: Masson.
(1896/1980). *Fragment d'histoire future*. Paris: Slatkine; Geneva: Ressources. (English transl.: *Underground man*. Foreword by H. G. Wells. New York: Macmillan, 1899.)

(1898a). *Les lois sociales – Esquisse d'une sociologie*. Paris: Alcan. (English transl. *Social laws*. Foreword by J. M. Baldwin. New York/London: Macmillan, 1899.)
(1898b). *Études de psychologie sociale*. Paris: Giard & Brière.
(1901/1904). *L'opinion et la foule*. Paris: Alcan.
(1973). *Écrits de psychologie sociale*. Toulouse: Privat.
Thayer, J. A. (1964). *Italy and the Great War – Politics and culture, 1870–1915*. Madison/Milwaukee: Univ. of Wisconsin Press.
Théveney, Général. (1932). À la mémoire du médecin inspecteur général Fournial. *La France Militaire*, Sept. 14, p. 3.
Thiec, Y. J. (1981). Gustave Le Bon – Prophète de l'irrationalisme de masse. In Besnard, ed.
Thiec, Y., and J. R. Théanton. (1983). La foule comme objet de "science." *Revue Française de Sociologie*, 24 (No. 1): 119–36.
Thomson, D. (1977). *Democracy in France since 1870*. Oxford: Oxford Univ. Press.
Tilly, C., L. Tilly, and R. Tilly. (1975). *The rebellious century, 1830–1930*. Cambridge, Mass.: Harvard Univ. Press.
Tommissen, P. (1967). Vilfredo Pareto als economist en socioloog. 2 vols., unpublished mimeo. Vilvoorde-Koningsloo.
Turner, R., and L. Killian. (1957/1972). *Collective behavior*. Englewood Cliffs, N.J.: Prentice-Hall.
Valbert, G. (1892). La théorie d'un positiviste italien sur les foules criminelles. *Revue des Deux Mondes*, Nov., pp. 202–13.
Valera, P. (1901/1973). *La folla – Romanzo*. Naples: Guida.
(1973). *Antologia della rivista "La Folla."* Naples: Guida.
Van Den Heuvel, G. (1987). De meute en het recht – Honderd jaar Scipio Sighele. *Intermediair*, 23 (No. 40; Oct. 2): 45–7, 63.
Van Ginneken, J. (1974). Het socialisme als massabeweging. *De Groene*, May 1.
(1982a). Macht over de massa. *Intermediair*, March 12, pp. 23–7.
(1982b). Een onderschatte Franse massabeweging. *Maatstaf*, No 6, pp. 61–73.
(1983a). Over de zogenoemde politieke psychologie en haar achtergronden. *Psychologie en Maatschappij*, No. 22, pp. 101–23.
(1983b). Gabriel Tarde en de ontdekking van de sociale psychologie. *Sociologisch Tijdschrift*, 10 (No. 1): 125–46.
(1984a). Massenpsychologie. In H. Lück, R. Miller, W. Rechtien, eds., *Geschichte der Psychologie*. Munich: Urban and Schwarzenberg, pp. 68–74.
(1984b). Le Bon, Tarde, Collective behavior and political psychology. In A. and J. Kuper, eds., *The social science encyclopaedia*. London: Routledge & Kegan Paul.

(1984c). The killing of the father – The backgrounds of Freud's Group Psychology. *Political Psychology,* 5 (No. 3): 391–414. (German transl.: Die Vatertötung – Über die Hintergründe von Freud's Massenpsychologie. *Psyche – Zeitschrift für Psychoanalyse,* 38 (No. 12): 1124–48.)

(1985a). The Italian origins of crowd psychology. In S. Bem, H. Rappard, W. Van Hoorn, eds., *Studies in the history of psychology and the social sciences.*

(1985b). The 1895 debate on the origins of crowd psychology. *Journal of the History of the Behavioral Sciences,* 21 (No. 4): 375–85. (German transl.: Die Diskussion von 1895 über die Ursprünge der Massenpsychologie. *Gruppendynamik – Zeitschrift für angewandte Sozialpsychologie,* 16 (No. 2): 85–94.)

(1985c). Rebellious mobs in 19th century European fiction. Paper presented at the 8th Annual Scientific Meeting of the International Society of Political Psychology. Georgetown Univ., Washington, D.C., June, 29 pp.

(1986a). Die Dritte Republik und die vernachlässigten Französischen Wurzeln der politischen Psychologie. In A. Schorr, ed., *Bericht über den 13. Kongress für angewandte Psychologie.* Vol. I. Bonn: Deutscher Psychologen Verlag, pp. 224–31.

(1986b). Tarde, Dreyfus and public opinion. In S. Bem, H. Rappard, and W. van Hoorn, eds., *Studies in the History of Psychology and the Social Sciences.* Vol. 4.

(1987a). Trotter, the herd instinct and the transformation of British liberalism. Paper presented at the 6th Annual Scientific Meeting of the European Cheiron Society for the History of the Behavioral and Social Sciences, University of Sussex, Falmer/Brighton (U.K.), 2–6 Sept., 25 pp.

(1987b). Reich's "Mass psychology of fascism" in Marxist perspective, *Revista de Historia de la Psicologia.*

(1988). Outline of a cultural history of political psychology. Ch. 1, pp. 3–22, in W. F. Stone and P. Schaffner, *The psychology of politics,* 2d ed. (compl. rev.). New York: Springer.

(1989a). De constructie van de mythe van de eenzame held – Het geval Le Bon (Review of Moscovici, Rouvier, and Vlach). *Psychologie en Maatschappij,* No. 48, pp. 253–65.

(1989b). Ten myths about Lombroso. *Cheiron Newsletter,* Autumn, pp. 26–30.

(1989c). Freud and the collapse of the Austro-Hungarian Empire, *Storia della Psicologia,* n.s., No. 1.

Van Ginneken, J., and R. Kouijzer, eds. (1986). *Politieke psychologie – Inleiding en overzicht.* Alphen a.d. Rijn: Samson.

Van Ginneken, J., and J. Jansz, eds. (1986). *Psychologische praktijken – Een twintigste eeuwse geschiedenis*. Den Haag: Vuga.

Van Heerikhuizen, B. (1973). Het eigenaardige van de sociologie – Durkheim en Tarde over de relatieve autonomie van het sociologisch vakgebied. Amsterdam: Sociology Inst./Univ. of Amsterdam (doct. scr./master's thesis, unpubl.).

Vaussard, M. (1961). *De Pétrarque à Mussolini – Évolution du sentiment nationaliste italien*. Paris: Armand Colin.

(1950, 1972). *Histoire de l'Italie moderne*. Paris: Hachette.

Vecchini, F. (1968). *La pensée politique de Gaetano Mosca et ses différentes adaptations au cours du XXe siècle en Italie*. Paris: Cujas.

Verbeek, Th. (1977). *Inleiding tot de geschiedenis van de psychologie*. Utrecht: Spectrum; Antwerp: Aula.

Vlach, C. (1982). "Sociologie et lecture de l'histoire chez Gustave Le Bon." Paris: Maison des Sciences de l'Homme (Thèse de 3me cycle).

Wallas, G. (1914/1936). *The great society – A psychological analysis*. New York: Macmillan.

Wertheimer, M. (1970). *A brief history of psychology*. New York: Holt, Rinehart & Winston.

Westermarck, E. (1927). *Memories of my life*. London: Allen & Unwin.

Whyte, A. J. (1965). *The evolution of modern Italy*. New York: Norton.

Widener, A., ed. (1979). *Gustave Le Bon – The man and his works*. Indianapolis: Liberty Press.

Woolf, H. (1961). *Quantification*. Indianapolis: Bobbs Merrill.

Worms, R. (1899). Psychologie collective et psychologie individuelle. *Revue Internationale de Sociologie*, April: 249 a.f.

Wright, H. M. (1961). *The new imperialism – Analysis of late nineteenth century expansion*. Lexington, Mass.: Heath.

Zeldin, T. (1979–81). *France 1848–1945*. 5 vols. Oxford: Oxford Univ. Press.

Zévaès, A. (1932). *L'affaire Dreyfus*. Paris: Éd de la Nouvelle Revue Critique.

Zimbardo, P. (1969). The human choice: Individuation, reason and order versus Deindividuation, impulse and chaos. *Nebraska Symposium on Motivation*, 17: 237–307.

Zimmermann, L. (1971). *Der Imperialismus*. Stuttgart: Klett.

Zola, E. (1960). *L'affaire Dreyfus – La vérité en marche*. Paris: Garnier-Flammarion.

Index

Note: In general, I have excluded lesser names and subjects mentioned only in passing and references to authors of recent secondary literature, with the exception of some that are particularly relevant.

Printed in the United States
68184LVS00002B/280

9 780521 032490